Lecture Notes in Electrical Engineering   **385**

More information about this series at http://www.springer.com/series/7818

Rolf Drechsler • Robert Wille
Editors

Languages, Design Methods, and Tools for Electronic System Design

Selected Contributions from FDL 2015

 Springer

Editors
Rolf Drechsler
University of Bremen
Bremen, Germany

Robert Wille
Johannes Kepler University
Linz, Austria

ISSN 1876-1100   ISSN 1876-1119   (electronic)
Lecture Notes in Electrical Engineering
ISBN 978-3-319-81106-2   ISBN 978-3-319-31723-6   (eBook)
DOI 10.1007/978-3-319-31723-6

Printed on acid-free paper

This Springer imprint is published by Springer Nature
The registered company is Springer International Publishing AG Switzerland

Foreword

The increasing integration and complexity of electronic system design requires a constant evolution of the used languages as well as associated design methods and tools. The *Forum on specification & Design Languages* (FDL) is a well-established international forum devoted to the dissemination of research results, practical experiences, and new ideas in the application of specification, design, and verification languages. It considers description means for the design, modelling, and verification of integrated circuits, complex hardware/software embedded systems, and mixed-technology systems.

FDL is the main platform to present and discuss new trends as well as recent work in this domain. Thanks to the commitment of the authors and presenters, also the 2015 edition of FDL was an interesting and lively meeting.

This book is devoted to FDL 2015 and contains the papers that have been evaluated best by both the members of the program committee and the participants of the forum which took place in September 2015 in Barcelona, Spain. It reflects thereby the wide range of topics which have been covered at this event. We categorized them into four major fields, namely:

1. Co-simulation for Automotive Systems
2. Reconfigurable Systems and FPGAs
3. Clocks and Temporal Issues
4. AMS Circuits and Systems

This portfolio provides an in-depth view on the current developments in our domain which surely will have a significant impact in the future.

As PC chairs of FDL 2015, we would like to thank all authors for their contributions as well as the members of the program committee and the external reviewers for their hard work in evaluating the submissions. Special thanks go to Julio Medina who served as general chair for FDL 2015 as well as Sophie Cerisier

and Adam Morawiec from the *Electronic Chips & Systems design Initiative* (ECSI).
Finally, we would like to thank Springer for making this book possible.

Bremen, Germany   Rolf Drechsler
Linz, Austria   Robert Wille
February 2016

Contents

Part I
Co-simulation for Automotive Systems

Chapter 1
Virtual Hardware-in-the-Loop Co-simulation for Multi-domain Automotive Systems via the Functional Mock-Up Interface

Ròbert Lajos Bücs, Luis Murillo, Ekaterina Korotcenko, Gaurav Dugge, Rainer Leupers, Gerd Ascheid, Andreas Ropers, Markus Wedler, and Andreas Hoffmann

1.1 Introduction

In the last several decades vehicles underwent a huge change from being mainly mechanical machines to digital hardware(HW)/software(SW)-centric vehicular systems. The evolution of automotive HW and SW in the last years shows an exponentially exploding complexity [9]. In this process vehicular SW has become highly diverse, ranging from infotainment to safety-critical real-time control applications [26]. With *Advanced Driver Assistance Systems* (ADAS) targeting autonomous driving scenarios, a high-end car of the last decade contained an estimated 100 million lines of source code [11]. The evolution of automotive HW, on the other side, shows an increasing trend towards more integrated architectures, i.e., fewer *Electronic Control Units* (ECUs) with multiple powerful processors. In today's high-end cars this results in around 100 networked ECUs consisting of almost 250 embedded and graphic processors in total [16].

Frequent factory recalls have been noted for several *Original Equipment Manufacturers* (OEM) in the last decade, where failures could have had severe effects on passenger safety [28]. For this reason, HW/SW functional security has become the number one priority for automotive systems. To overcome insufficient HW/SW test

R.L. Bücs (✉) • L.G. Murillo • G. Dugge • R. Leupers • G. Ascheid
Institute for Communication Technologies and Embedded Systems,
RWTH Aachen University, Aachen, Germany
e-mail: buecs@ice.rwth-aachen.de; murillo@ice.rwth-aachen.de; dugge@ice.rwth-aachen.de; leupers@ice.rwth-aachen.de; ascheid@ice.rwth-aachen.de

E. Korotcenko • A. Ropers • M. Wedler • A. Hoffmann
Synopsys Inc., Aachen, Germany
e-mail: korotcenko@synopsys.com; ropers@synopsys.com; wedler@synopsys.com; hoffmann@synopsys.com

© Springer International Publishing Switzerland 2016   3
R. Drechsler, R. Wille (eds.), *Languages, Design Methods, and Tools for Electronic System Design*, Lecture Notes in Electrical Engineering 385, DOI 10.1007/978-3-319-31723-6_1

and validation techniques, new mandatory functional safety standards have arisen, with the *ISO26262* [17] among them, targeting full correctness of the electrical subsystem. Although these requirements need to be mandatory to ensure functional safety for vehicles, they also profoundly hamper HW/SW system development. Needless to say, with exponentially rising amount of SW, emerging complex multi- and many-core technologies and demanding requirements posed by functional safety standards, automotive HW/SW system development and validation have become immensely difficult.

Virtual Platform Technology A promising solution to tackle the aforementioned challenges is *virtual platform technology* [30]. A *virtual platform* (VP) is the ensemble of simulation models representing HW blocks and their interconnection. VPs are usually assembled using high-level programming languages that also enable modeling and simulation such as *SystemC/TLM2* [35]. VPs can be used to comprehensively simulate a target architecture as well as to model the system at a higher abstraction level, while still maintaining an accurate view for SW developers. VPs also have the advantage of providing full HW/SW visibility and controllability without altering the simulation. In addition, the HW/SW simulation can be repeatedly executed with exactly the same behavior if persistent inputs are provided. These properties facilitate debugging, testing and validation efforts of the real HW/SW systems to be developed. Thus, VP technology is a promising way to decrease design complexity and to bridge the verification and functional safety gap of vehicle's electrical systems. But because vehicles are highly heterogeneous, full-vehicle simulation requires to capture the interactions of individual subsystems (domains) since strong inter-dependencies exist between them. This limits the usage of virtual platforms alone and calls for modeling and simulation techniques *beyond* domain boundaries.

Multi-domain Co-simulation via FMI *Multi-domain co-simulation* enables to comprehensively interconnect all involved domains of a heterogeneous system, and thus sheds light on previously neglected inter-domain influence. For example, in case of the simulation of an electric vehicle's power connector: the electrical, mechanical, and thermal properties of the device influence one another significantly, and thus have to be considered together. Similarly, the behavior of ADAS appli-cations is not just influenced by the HW/SW systems embedding them but also physical inputs coming from the environment (e.g., temperature, road properties, atmospheric conditions), captured by peripheral devices (e.g., sensors).

To achieve multi-domain simulation, several simulation interoperability stan-dards, tools and approaches have emerged, with the *Functional Mock-Up Interface* (FMI) as the most promising among them. FMI is an open-source, tool-independent standard supporting model exchange and co-simulation [14]. It was originally initiated by Daimler, and it is since then strongly driven by the automotive industry with the primary goal of exchanging simulation models between suppliers and OEMs. Recently, the goal of FMI has become much more general: to support a fully simulation-based development and testing environment of complex virtual devices consisting of many coupled subsystem models from different vendors. This approach has made FMI the de facto co-simulation standard for automotive.

To integrate the HW/SW system into automotive multi-domain simulation, it is essential to transform VPs into FMI-compliant modules, also called *Functional Mock-Up Units* (FMUs). This requires not just connecting the event-driven SystemC simulation approach with the mainly continuous-time FMI concept, but also to ensure, that *Electronic System Level* (ESL) modules can fulfill the FMI specification.

Contribution This chapter is based on the work presented in [10], and the contributions are as follows:

  (i) As to the best knowledge of the authors, no attempts have yet been made to create virtual platforms as FMUs. Thus, two methods will be demonstrated to integrate SystemC-based VPs into heterogeneous multi-domain automotive simulation systems via the FMI standard.
 (ii) Both approaches have been applied on state-of-the-art ECU VPs to achieve *virtual Hardware-in-the-Loop* (vHIL) simulation.
(iii) The *Parallel Functional Mock-Up Interface* (PFMI) master will be presented, which is an advanced FMI simulation controller with features to orchestrate the co-simulation, facilitate system assembly and evaluation, as well to ensure system robustness via system-level fault injection capabilities.
 (iv) The advantages of the achieved multi-domain simulation will be evaluated by executing safety-critical ADAS algorithms on the instrumented VPs, and analyzing their HW/SW functionality via practical driving scenarios.
  (v) The advantages of system-level fault injection will be demonstrated so to increase functional safety testing and overall HW/SW system robustness.

Outline The rest of this chapter is structured as follows. The main simulation principles of FMI and SystemC, and their possible interconnection, is discussed in Sect. 1.2. Afterwards, Sect. 1.3 presents two possible methods to create VP FMUs. Subsequently, Sect. 1.4 describes the PFMI master. Performance measurements, the evaluation of the PFMI master, as well as both applied VP FMU approaches are demonstrated in Sect. 1.5. Related scientific contributions are evaluated in Sect. 1.6. Finally, Sect. 1.7 draws the conclusions of the work.

1.2   Background

In order to successfully interleave the FMI and SystemC worlds it is essential to understand the simulation principles behind them so to detect and bridge conceptual gaps. The key aspects are described as follows.

General Simulation Principles The FMI specification defines a master/slave simulation concept, where the *FMI master* is a single module responsible for synchronization, communication, and data exchange between the involved slaves.

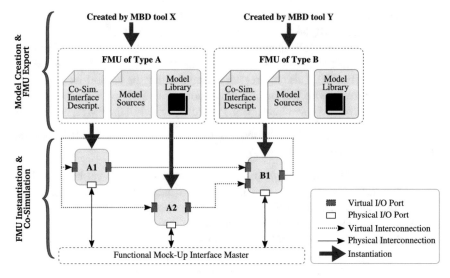

Fig. 1.1 Abstract example for FMU creation, export, interconnection, and co-simulation

FMI slaves, or *Functional Mock-Up Units* (FMUs), are modules encapsulating a certain co-simulation model and implement the lightweight FMI C-API [14].

To visualize the general principles behind FMI, Fig. 1.1 depicts an abstract example. Here a top-down design flow is assumed, i.e., *Model-Based Design* (MBD) tools are used to create high-level simulation models of physical systems. The development step includes system prototyping and functional testing within the tool. In this example, two such frameworks (tool X and Y) are shown, that are also capable of an FMI-compliant simulation model generation (in form of FMUs). Furthermore, in this example two types of such models (FMU of type A and B), and three model instances (A→A1, A→A2, B→B1) are created. The goal here is to be able to involve models of the previously mentioned, not interconnected MBD tools in a co-simulation for, e.g., further model integration steps or system testing. To interconnect them, the system designer can determine connections between instantiated FMUs. These connections, and the input/output model ports, are denoted as *virtual*. The reason behind this is, that actual *physical* connections are handled and I/O data is forwarded through the FMI master only. The FMI API also defines further responsibilities of the master module (e.g., synchronization, control), but does not specify how to implement these explicitly. Such responsibilities of both FMI master and slaves, as well as structural considerations, are described in the following.

Coupling Structure Regarding their content and coupling structure, FMUs can be created either as *tool-coupling* or *standalone* modules, as depicted in Fig. 1.2. Following the first approach (Fig. 1.2a) the FMU does not contain the simulation execution engine (kernel or solver) which, thus, has to be present on the simulation host. Here, the FMU itself is a generic communication module (e.g., an

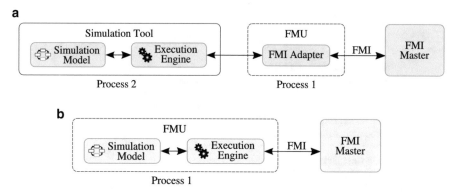

Fig. 1.2 Tool-coupling vs. standalone FMU. (**a**) Tool-coupling FMU structure. (**b**) Standalone
FMU structure

FMI adaptor) interacting with the simulation tool which contains the model and
the simulation execution engine. This concept is used by models of proprietary
simulators.

Following the second approach (Fig. 1.2b) the FMU is self-contained and
encapsulates the model, the simulation execution engine, as well as all required
runtime dependencies, and can be thus executed autonomously. In this chapter, we
elaborate on solutions proposed for VPs based on both coupling structures.

Model Content An actual FMU is a bundle of files containing the following:

- *Model Description File:* describes basic properties of an FMU and the definition
  of all exposed variables, i.e., the external interface. An example model description XML is depicted in Fig. 1.3. As seen here, this file also describes several
  co-simulation capabilities of the FMU (e.g., simulation step information, tool-
  coupling/standalone model).
- *FMU Model Implementation:* in form of source code and/or pre-compiled shared
  libraries.
- *Additional Files:* runtime library dependencies, model resources, documentation,
  model icon, among others.

In contrast, SystemC/TLM2-based VPs consist of the hierarchical description of
HW modules and their interconnection in form of C/C++ source files and/or model
libraries, as well as the SW stack to be loaded for each core within the VP. Further
runtime library dependencies and the SystemC kernel have to be available on the
simulation host.

Simulation Entry/Exit According to the standard, FMUs are created and
destroyed *explicitly* via separate dedicated functions. Furthermore, the initialization
and termination of an FMU is also distributed through multiple intermediate steps,
each encapsulated in an API function. The calling sequence of these procedures
(also depicted in Fig. 1.4a) is as follows:

```
<fmiModelDescription
  fmiVersion="2.0"
  guid="de739a83−1570−41a7−9f4d−55c19d96b956"
  modelName="ARM_VP"
  description="ARM_FMU">

  <CoSimulation
  canHandleVariableCommunicationStepSize="true"
  modelIdentifier="ARM_VP"
  needsExecutionTool="true"/>

  <LogCategories>
    <Category name="logAll"/>
  </LogCategories>

  <ModelVariables>
    <ScalarVariable causality="output" name="IRQ" valueReference="0">
      <Boolean/>
    </ScalarVariable>
  </ModelVariables>

</fmiModelDescription>
```

Fig. 1.3 Example FMU model description XML

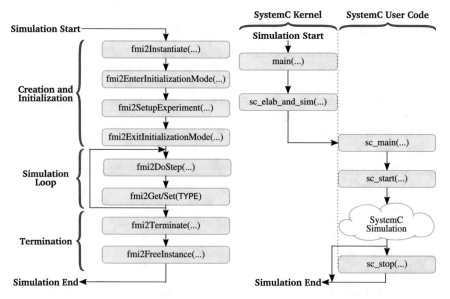

Fig. 1.4 FMI vs. SystemC simulation flow. (**a**) FMI simulation flow. (**b**) SystemC simulation flow

- `fmi2Instantiate(...)`: creates an FMU instance by specifying its name, type, unique identifier, resources, callback functions, and logging properties.
- `fmi2EnterInitializationMode(...)`: notifies the FMU to perform its encapsulated model's initialization.
- `fmi2SetupExperiment(...)`: triggers the setup of the model's boundary conditions (if any) and fixes the simulation time span ($t_{start} \rightarrow t_{end}$).
- `fmi2ExitInitializationMode(...)`: instructs the FMU to exit the previously invoked initialization mode.
- `fmi2Terminate(...)`: notifies the FMU that the simulation run has been terminated (e.g., explicitly or the simulation end time has been exceeded).
- `fmi2FreeInstance(...)`: frees the allocated memory and other reserved resources, and unloads the FMU.

In SystemC these actions are carried out *implicitly* when executing the simulation, i.e., by calling the `main(...)` entry point of the kernel. The nested function call chain (also depicted in Fig. 1.4b) is as follows:

- `sc_elab_and_sim(...)`: initializes the SystemC kernel's internal state.
- `sc_main(...)`: the entry point of the actual SystemC simulation. This function (either *implicitly* or *explicitly*) allocates the SystemC modules, signals, I/O ports, and further objects, and performs their interconnection.
- `sc_start(...)`: executes the SystemC simulation mostly in its complete duration (specified by the user).
- `sc_stop(...)`: explicitly terminates the simulation run.

After the simulation has been finished, `sc_main(...)` returns inferring the cleanup of all SystemC constructs and the synchronization of the kernel's state.

Simulation Run-Control A significant conceptual difference between modeling with FMI and SystemC lies in the continuous-time vs. event-driven execution mechanisms. FMI has been tailored for continuous-state systems described by ordinary differential equations. These are solved by numerical integration methods as simulation time advances by either a fixed or variable time step. Following the FMI API, the execution of a slave (until a given time) needs to be triggered by the `fmi2DoStep(...)` function. As shown in Fig. 1.4a, this function is called periodically by the FMI master until the simulation time ends.

SystemC, in contrast, supports an event-driven simulation principle, where only certain phenomena change the state of the simulation at discrete points in time which are not defined in advance. Simulation time advances across discrete timestamps, where pending events are waiting to be processed.

Inter-Module Communication Data exchange between FMUs can be achieved via the type-specific set of `fmi2Get/Set(TYPE)` functions. These are triggered by the FMI master between simulation steps if an FMU is connected to another. The master first fetches an output value by calling the `fmi2Get(TYPE)` function of the origin, and then propagates this value to the target by calling it's `fmi2Set(TYPE)` function. Thus, data exchange between FMUs happens always through the master.

In contrast, modules within two separate SystemC simulations cannot communicate with each other. However, this is possible within one simulation via multiple SystemC constructs such as signals, ports, exports, events, channels, and interfaces.

1.3 Methodology

1.3.1 Tool-Coupling VP FMU Concept and Creation

Taking the aforementioned conceptual differences in account, and following the tool-coupling VP FMU scheme (Fig. 1.2a), the idea of a *SystemC FMU Adaptor* has been proposed. This central communication component implements all FMI C-API functions and interconnects them with SystemC constructs as follows:

- It prepares the VP for simulation at instantiation time. This includes:

  – Allocating memory for the VP, if necessary.
  – Exposing the specified I/O interconnects of the VP to the FMI master and thus to the co-simulation.
  – Initializing the SystemC kernel's internal state.
  – Setting the SystemC time base and fast-forwarding the simulation to the first simulation delta cycle.

- To correctly execute the whole simulation, it drives the VP to execute simulation and communication intervals one by one. The former can be achieved by invoking sc_start(...) with the stepsize requested by the FMI master.
- The module provides a data exchange mechanism between the VP and the FMI master. Internally, the fmi2Get/Set(TYPE) functions need to be connected to entities, that can access the I/O of the VP.
- Finally, sc_stop(...) needs to be called to terminate the simulation run, and all allocated SystemC objects need to be cleaned up explicitly.

To put these concepts into practice, a generic tool-coupling VP FMU generation framework has been developed. The framework is based on *Platform Architect* (PA), a commercial tool for assembling SystemC-based VPs, provided by *Synopsys* [33]. The primary goal was to achieve a one-click SystemC FMU Adaptor generation from within PA, after modeling, assembling, and determining co-simulation properties of a VP. This would also allow the system designer to use advanced debugging and simulation control facilities of the Synopsys infrastructure.

To achieve this, interconnection possibilities and the platform creation and export flow of PA needed to be examined. For co-simulation purposes, the Synopsys infrastructure provides the *Virtualizer System Interface* (VSI) API [34], which can be used with VPs created in PA. This library provides mechanisms for synchronization and data exchange between Synopsys VPs and a few external simulation tools (e.g., Simulink). Data exchange can be achieved via VSI *connector blocks*, which have

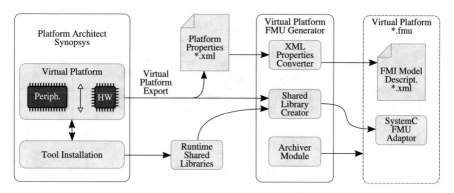

Fig. 1.5 Tool-coupling VP FMU generation framework

to be instantiated both in the VP and the external model of simulation. Thus, VSI is rather a point-to-point coupling solution, but its features have the potential to be re-used to achieve a flexible, generic co-simulation (e.g., through FMI).

An abstract block diagram of the VP export flow and the tool-coupling VP FMU generation framework is depicted in Fig. 1.5. After loading or assembling a platform in PA, the designer needs to expose its I/O ports so to be able to use them in a co-simulation environment. This can be achieved by instantiating VSI connector blocks and connecting them to signals or ports that shall be externally accessible. Afterwards, the complete virtual HW/SW system can be exported into one simulation binary. This procedure also generates a platform description file containing essential information about the VP, such as hierarchy description of all HW components in the platform, as well as their interconnection. To create the FMU model description XML, exactly these attributes are extracted via the *XML Properties Converter* module of the FMU generation framework. The resulting file contains essential information about the tool-coupling VP FMU, as well as co-simulation properties, that have to be known by the co-simulation master. This includes name, data type, and direction of the instantiated VSI I/O blocks, model variables, whether the FMU can handle a variable stepsize, whether it needs an external execution tool, or whether it can roll back a simulation step, among many others. Finally, the *Shared Library Creator* will generate the SystemC FMU Adaptor, which will act as the tool-coupling module between the simulation tool executing the model and the FMI master.

The presented method and the created framework enable a fast and convenient VP FMU creation. In general, the tool-coupling method is suitable for co-simulation models with proprietary simulators. The limitation of the approach is I/O handling with connector blocks, mostly because it needs extra modules to be added to the VP, that would not be included otherwise. These extra blocks induce extra events to be processed by the SystemC kernel, which might also affect simulation performance.

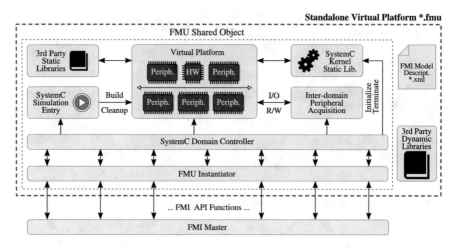

Fig. 1.6 Standalone FMU internal structure and interconnection scheme

1.3.2 Standalone VP FMU Concept and Creation

As an alternative approach, guidelines for creating standalone VP FMUs have been also created. The goal was herewith, that the created modules do not require specific toolchains to be available on the simulation host, and thus can be distributed and co-simulated with minimal integration effort. This highly facilitates the integration of full ESL HW/SW systems into more complex virtual devices, such as a heterogeneous full vehicle model.

To overcome the main conceptual and structural differences between FMI and SystemC (described in Sect. 1.2), the standalone VP FMU approach proposes the following modules (also depicted in Fig. 1.6). The first essential component is the *SystemC Simulation Entry*, responsible for dynamically allocating/deallocating all SystemC constructs at creation/destruction time. It is also responsible of connecting signals and ports at initialization time (if this is not done in a top module entity explicitly). Furthermore, the previous FMU adaptor approach has been replaced with a *SystemC Domain Controller* responsible for implementing and applying the FMU API calls (from the FMI master) on the SystemC simulation as follows:

- Initialize the SystemC kernel and request the construction of SystemC objects.
- After initialization, fast forward simulation to the first delta cycle by calling `sc_start(SC_ZERO_TIME)`.
- Achieve a step-based simulation run-control with a constrained step-time (requested by the FMI master) via periodic calls to `sc_start (fmu_steptime)`.
- Implement I/O exchange between the FMI master (and thus other FMUs) and peripherals of the VP.

- Apply reset, if requested by reinitializing the SystemC kernel's state and resetting all SystemC components.
- If the end of simulation has not yet been invoked, terminate the simulation by calling `sc_stop(...)` explicitly.
- Request the destruction of SystemC constructs and stop the kernel.

Generic Inter-Domain Communication   The previous approach, using connector blocks, has been replaced, so that a VP does not need to be extended with additional SystemC modules. The new approach relies on an *Inter-Domain Peripheral Acquisition* component, that automatically registers chosen platform peripherals to the SystemC Domain Controller. This mechanism directly enables the peripheral's ports to be exposed to the FMI master (and thus to other FMUs). This approach is also highly flexible to externally access peripherals within a VP, and might even be applied for third-party models. Furthermore, the system designer can implement the I/O buffering strategy of choice. In this work two methods have been implemented:

- FIFO input and output buffers with a user-defined size.
- A single value, following the 'last is best' approach used in reactive control systems, such as the ones in vehicles.

Structural Considerations   To add modularity to the FMI API function stack, the *FMI Instantiator* component has been proposed. This module separates generic from FMU-specific functionality. Former services are handled in the same way for different kinds of FMUs (e.g., setup debugging log, allocate/deallocate memory, status requests, handle enter/exit to/from simulation modes). The FMU-specific functionality is handled then by the particular SystemC Domain Controller individually.

The last requirement for a VP to undergo the standalone FMU transformation is to resolve model dependencies internally. According to the FMI standard, multiple instances of the same FMU are allowed to be loaded and used by the FMI master at the same time, which can lead to certain pitfalls. It can happen, for example, that these identical instances share the usage of dynamic libraries, e.g, the SystemC kernel. Thus, the internal models of these FMU instances, possibly in different states, would use the same kernel and further dependent library instances, which might cause inconsistency in their state. Needless to say, such situations can lead to malicious errors at runtime. The best practice according to our experience is to (1) apply static library linkage and to (2) bind symbol references to the global definitions within their libraries. However, this solution is not applicable with third-party runtime dependencies. To achieve a completely self-contained standalone FMU, these libraries have to be inserted into the FMU bundle. Then the FMI master can individually load these dynamic dependencies and, in case of multiple instances, it might also need to copy and rename them for complete isolation. A further solution to the aforementioned problem is to start all FMU instances in a separate execution context. Only this leads to a complete and safe separation of the FMU instances with the trade-off of a significantly higher inter-FMU communication latency.

1.4   The Parallel FMI Master

The proposed PFMI master, was tailored focusing on simulation performance, usability, user-friendliness, and most importantly, full-system evaluation capability, and functional safety. The general objective was to support standalone FMUs. Thus, on one hand, the master applies a simple fixed-step master algorithm, assuming that the kernel or solver, contained by the FMU, further refines the substeps, if necessary. On the other hand, PFMI supports parallel simulation and I/O propagation capabilities. The system-level fault injection feature of PFMI, together with the multi-domain nature of the co-simulated systems, can close the functional safety gap for the HW/SW system. PFMI's main components and functionalities, as shown in Fig. 1.7, will be described in the following.

1.4.1   FMU Import, Interconnection and
Co-simulation Control

In PFMI importing FMUs can be done in the GUI either by selecting and extracting an `*.fmu` bundle, or by selecting the model description file of an FMU explicitly. During loading, PFMI offers to setup an FMU's environment without affecting other FMUs. To prepare the selected FMUs for co-simulation, the *System Configuration Manager* will be triggered, which internally creates an *FMU Wrapper* object for each loaded FMU. These components process all information specified in the model description XML and encapsulate the actual simulation model. This includes implementing services and facilities to, e.g., allocate/deallocate memory for the model, check and set up the I/O, implement logging, initialize the model's environment, resolve runtime dependencies, and request the instantiation of the FMU. Loaded FMUs can be connected within PFMI either via the GUI or by loading a

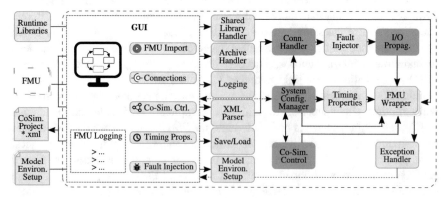

Fig. 1.7 Internal structure of the PFMI master

previously saved co-simulation scenario file. This activates the *Connection Handler* module internally, which checks port directions and data types, and registers the specified connection. The actual data exchange is happening via the *I/O Propagator* which triggers the `fmi2Get(TYPE)` of the origin and the `fmi2Set(TYPE)` of the target port between every simulation interval. After loading, initializing and instantiating the FMUs, the actual co-simulation can be executed. PFMI provides means for co-simulation run-control, thus the user can either step, continuously run, or stop the simulation execution. Internally, these operations trigger the *Co-Simulation Control* component, which is responsible for collective execution control of all FMUs. Finally, it is worth to mention, that I/O propagation as well as step mechanisms, can be selected to be sequential or parallel (the latter being supported via *OpenMP* [25]).

1.4.2 Automated Step Constraint Resolution

In PFMI, the user is able to manually select a collective fixed stepsize for all FMUs. Internally, this value will be propagated by the *Co-Simulation Control* to all *FMU Wrappers* and thus to all models. More importantly, PFMI also supports a semi-automated optimal fixed stepsize identification. This is essential for FMUs with time step constraints, e.g., with kernels defining a maximum lookahead time, such as the one presented in [38]. After entering the step constraints and minimal resolution of the interconnected FMUs, the *Timing Properties* module can determine the optimal stepsize via the identification algorithm shown in Algorithm 1 as follows.

First, PFMI determines the interconnection scheme of the co-simulation participants by automatically assembling the FMU connection graph. Afterwards, it iterates over all connections with the goal to identify *FMU timing clusters*, which are sets of connected FMUs within the connection graph. Co-simulation modules within a cluster will share either the same or a multiple of an agreed base stepsize. Afterwards, it iterates through all identified clusters and FMUs within them to fetch the minimum and maximum supported step size. Thus, it can identify the steprange of individual FMUs and can calculate all possible stepsizes via the provided time resolution. Subsequently, the algorithm iterates through all clusters and tries to intersect the possible step values. The goal is to find one that is acceptable for all FMUs within a cluster, if this even exists. Finally, if more possible stepsizes have been found, the maximum will be chosen, so that the kernel overhead of stepping can be minimized (i.e., less steps are required).

1.4.3 System-Level Fault Injection Capabilities

The motivation for fault injection in a vehicular multi-domain simulation is to increase system robustness. This feature has the ability to close the functional

---

Algorithm 1 Optimal fixed stepsize identification

 1: **for** *c* in connection graph **do**
 2:     **for** *cli* in set of clusters **do**
 3:         *oidx* = *find* origin index of *c* in *cli*;
 4:         *tidx* = *find* target index of *c* in *cli*;
 5:         **if** *oidx* found **then**
 6:             Add *tidx* to *cli*
 7:         **else if** *tidx* found **then**
 8:             Add *oidx* to *cli*
 9:         **else**
10:             Create new cluster, add *oidx* and *tidx*
11:         **end if**
12:     **end for**
13: **end for**
14: **for** *cli* in set of clusters **do**
15:     **for** *fmu* in *cli* **do**
16:         Get supported *steprange*
17:         Calculate all possible *stepsizes* for *fmu*
18:     **end for**
19:     intersect all supported *stepsizes* in *cli*
20:     select max(*stepsize*) as cluster step
21: **end for**

---

safety gap of the virtual HW/SW system with a simulation-based evaluation, testing, and verification approach. Thus, following the guidelines described in [20] PFMI has been extended with fault injection features. While not violating the FMI specification, there is one effective way for a master to inject faults in a system: alter the I/O exchange of connected FMUs. In PFMI, the user can select FMU connections via the GUI, and insert a fault of the following types:

- *Link Broken*: I/O is not propagated at all on the specified connection.
- *Constant Additive Offset*: I/O is propagated with a user-defined offset on the specified connection.
- *Constant Multiplicative Offset*: I/O is propagated with a user-defined gain on the specified connection.
- *Stochastic Additive Noise*: I/O propagation with additive noise with a given statistic property and amplitude.

The fault type can also be augmented with its occurrence:

- *Permanent Fault*: happens in every communication cycle.
- *Periodic Fault*: happens in every nth communication cycle.
- *Spurious Fault*: happens with a specified probability of occurrence.

Internally, the *I/O Propagator* creates a *Fault Injector* module for every chosen connection, fault and occurrence type, that was previously configured by the user. The fault injection features of PFMI are one step towards system-level simulation-based unit testing, and will be evaluated in the following chapter.

1.5 Test Cases, Benchmarks and Evaluation

In order to evaluate the created framework and the applicability of the proposed approaches, several VP FMUs have been created and used in different practical multi-domain co-simulation scenarios together with other FMI-compliant components. PFMI was used as the co-simulation master, and with the standalone VP FMU use-case, fault injection experiments have been conducted so to increase robustness and facilitate functional safety testing of the HW/SW subsystem.

1.5.1 Simulation Performance Measurements with PFMI

To figure out which configurations are mostly profitable for parallel simulation, performance measurements have been carried out for PFMI executing a set of synthetic co-simulation benchmarks. The basic co-simulation scenario involves a source FMU, creating and sending data to a sink FMU, which processes the data and writes it into a file. This setup has been mirrored and grouped into a single co-simulation to achieve use-cases with more FMUs. Moreover, every experiment was configured with three different data sets (marked as #1, #2, and #3 in Table 1.1) resulting in a growing computational complexity of data generation and processing. All measurements have been conducted on an Intel Core i7 CPU 920 simulation host: quad-core, `fclk` =2.67 GHz, 2×128 KB L1 instruction/data, 1 MB L2, 8 MB L3 cache with 12 GB RAM. For each co-simulation run and configuration host cycle counts were captured. Table 1.1 shows the average host cycle expense of the following operations:

  (i) the FMU's `fmi2DoStep(...)` functions,
 (ii) the PFMI calling sequence to all FMUs in a sequential setting and
(iii) the PFMI calling sequence to all FMUs in a parallel setting.

As shown in Table 1.1, parallel simulation is beneficial if the execution of a particular FMU's step function requires more than around 120k host cycles (#1). The maximum speedup achieved with the synthetic benchmarks was 4.06x in a six FMU use-case, which can be explained with caching effects and optimal load balancing. The achievable speedup scales with the number of cores and the cache size of the simulation host.

Because of performance concerns, the overhead of PFMI, the FMU wrapping layer, and the cost of I/O propagation compared to the total step cost of the models in the previous test cases have also been profiled. For PFMI, a maximum of 0.18 % (with an average of 0.035 %) performance overhead has been measured. The FMU wrapper and I/O propagation resulted in an even lower overhead (in average 0.013 %). These results highlight how lightweight PFMI and the FMI layer are compared to the computational complexity of an actual simulation model.

Table 1.1 PFMI performance measurements ([kc]: kilo host cycles, [x]: times)

Benchmark data set	2 FMU			4 FMU			6 FMU		
	#1	#2	#3	#1	#2	#3	#1	#2	#3
FMU Average step cost [kc]	126.5	806.5	1206.4	126.5	806.5	1206.4	126.5	806.5	1206.4
Tot. sim. step cost seq. [kc]	253.1	1613.1	2412.9	506.2	3226.1	4826.1	759.4	4838.9	7239.2
Tot. sim. step cost par. [kc]	203.3	917.1	1338.1	359.4	1108.8	1508.4	549.1	1373.9	1781.7
Speedup par. vs. seq. [x]	**1.24**	**1.76**	**1.80**	**1.4**	**2.9**	**3.2**	**1.38**	**3.52**	**4.06**

To put the results into perspective of an automotive FMU, a quarter vehicle model (created using Simulink's *Simscape* [31] library) has also been profiled. This component includes engine, transmission, brake, and tire models, as well as the chassis block. The profiling results have shown an average step cost of `1282k` host cycles for this model, which is significantly higher than the step complexity of FMUs, where parallel simulation has been shown to achieve gains. Since our scenarios involve multiple FMUs with high step complexities, parallel simulation with PFMI is beneficial for these use-cases if a good load balancing is guaranteed.

1.5.2   Tool-Coupling VP FMU Approach

To evaluate the tool-coupling VP FMU generation framework, an expandable version of the Freescale Qorivva MPC5643L [15] ECU VP (provided by Synopsys) has been used. The platform comprises a 32-bit *Instruction-set-Accurate* (IA) dual-core PowerPC *instruction set simulator* (ISS), communication peripherals (e.g., FlexCAN and DMA), among many other modules detailed in [15]. This *virtual Electric Control Unit* (vECU) has been augmented with connector blocks within PA, exported for simulation, and finally our framework has been used to generate a tool-coupling VP FMU. The main goal was to achieve an initial co-simulation including this vECU and a basic vehicle model comprising of multi-domain physical components. For the latter, the Simulink *sf_car* model [32] has been chosen and tailored into a standalone FMU manually. The model is a simplistic quarter vehicle block consisting of an engine, transmission, and a chassis physical component. It also includes an automatic gear shift controller modeled via Simulink state machines, which was removed and exported as embedded code to be executed on the vECU. The quarter vehicle has been exported as a host-based simulation model via Simulink's code generation facilities and manually converted into an FMU. To provide the communication link, a basic SystemC/TLM2 *Controller Area Network* (CAN) transceiver model has been developed and transformed into an FMU. This component converts the physical I/O values of the vehicle model into CAN messages and vice versa to ensure correct communication with the vECU FMU. For debugging purposes, a complementary CAN logger FMU has been added to the co-simulation as well. The resulting co-simulation scenario is shown in Fig. 1.8.

The simulation results, as captured by the logger, are depicted in Fig. 1.9. Here, a strong acceleration maneuver is applied by increasing the throttle linearly from `55 %` to `90 %`. The automatic gear shift logic, extracted from the vehicle model and executed on the vECU, decides when to perform an up- or downshift based on the engine RPM, the current gear, and the vehicle speed, output by the vehicle model. As expected, the control SW shifts up at given points in time to achieve high reserves of engine torque at a particular RPM.

This evaluation demonstrates, firstly, functional and behavioral correspondence of the VP-based detailed and the host-based abstract simulation models. Secondly,

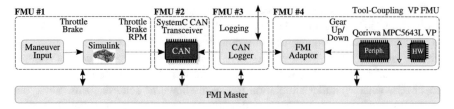

Fig. 1.8 Tool-coupling VP FMU co-simulation scenario

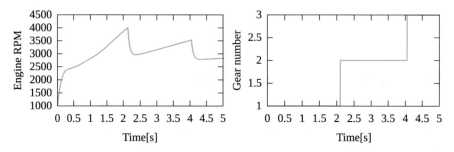

Fig. 1.9 Co-simulation results: tool-coupling VP FMU scenario

having the VP in the loop adds additional timing details to the actual ADAS algorithm implementation, which helps to anticipate how it would temporally behave on the real target. Finally, the scenario highlights the necessity of multi-domain simulation as (1) different simulation models of different physical domains affect each other's behavior and (2) such an immediate connection is an inevitable requirement for closed loop control systems.

1.5.3 Standalone VP FMU Approach

To test the standalone VP FMU approach, a novel embedded automotive SystemC TLM2-based VP, the *Virtual AUTOTILE Platform* (VAP), has been created. As illustrated in Fig. 1.10, VAP consists of a configurable number of uniform subsystems, also denoted as vECUs, or tiles, connected to and communicating via a CAN bus. Each subsystem contains an IA model of an in-house RISC processor based on *Dynamic Binary Translation* (DBT) simulation technology [19], an on-chip bus, an on-chip memory, a CAN transceiver, and miscellaneous peripherals (e.g., UART, timer, watchdog, interrupt controller, memory management unit). The VAP can be scaled to contain up to approximately 1000 tiles, thus enabling to model and simulate highly complex electronic systems expected in next generation cars. The platform has been converted into a standalone FMU following the steps described in Sect. 1.3.2. The CAN bus (as well as the UART of each tile) has been registered for the Inter-Domain Peripheral Acquisition module of the platform.

The VAP has been used within a top-down design flow for developing safety critical ADAS applications as the last step before using real hardware. The flow also includes algorithm prototyping and functional testing in Simulink, code generation to obtain embedded C code for a particular target, and code integration. VAP is employed during integration and system testing because it highly facilitates these steps compared to using real hardware. Following this design flow three ADAS algorithms have been created, which we wanted to evaluate using the VAP in a vHIL approach:

- *Anti-lock Brake System* (ABS): regulates the brake torque to keep wheel slip around an optimum when braking.
- *Traction Control System* (TCS): regulates the throttle to keep wheel slip around an optimum when accelerating.
- *Adaptive Cruise Control* (ACC): providing user inputs of desired speed and desired minimal distance from a leading vehicle, the algorithm keeps the set distance or the set speed, if no obstacle is present in front of the car.

Following the model-based design flow, after in-tool testing, these applications have been integrated as embedded C-code into three separate VAP subsystems (as depicted in Fig. 1.10). For the evaluation, a different quarter vehicle block has been chosen (based on Simulink's Simscape), described previously in Sect. 1.5.1. This model has been configured, exported as C/C++ code, and converted into a standalone FMU, also containing the solver.

In order to highlight the heterogeneity of the FMI approach, two other models have been designed via *OpenModelica* [24], an open-source MBD tool. Since this framework supports FMI model export, these modules could be directly exported as standalone FMUs. The first one is a user input block providing data to the control

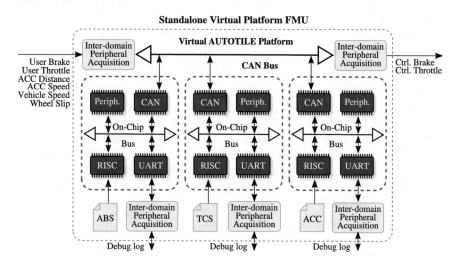

Fig. 1.10 The Virtual AUTOTILE Platform augmented with special modules for co-simulation

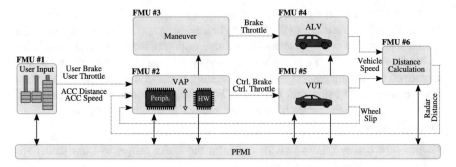

Fig. 1.11 Co-simulation scenario with the VAP standalone FMU

systems running on the VAP such as brake and throttle signals, desired distance, and speed for the ACC. The second is a module to calculate the distance of two vehicles based on their current speeds. These blocks are denoted as `FMU #1` and `FMU #6` in Fig. 1.11, depicting the FMU interconnection scheme.

1.5.3.1 Test Scenario: Driving on a Highway

A highway driving scenario has been defined as follows:

- The two virtual vehicles are initially placed 100 m from each other.
- The artificially maneuvered *Abstract Leading Vehicle* (ALV) model holds an initial speed of 75 km/h.
- The *Vehicle Under Test* (VUT) is initially standing still.
- The ACC of the VUT is set to a desired speed of 90 km/h and a desired minimal following distance of 50 m.
- At around 300 s, the ALV's speed is set to 110 km/h (abruptly).
- Finally, the ALV is controlled to apply steady braking until 60 km/h.

This setup mimics one car cruising and another entering a highway. The co-simulation results are depicted in Fig. 1.12a–c. After simulation start, the VUT accelerates up to the desired speed until it gets too close to the ALV, and the virtual distance measurement triggers the ACC brake control. Afterwards, as desired, the VUT cruises behind the ALV with its current speed, while keeping the desired distance. After the ALV accelerates up to 110 km/h the VUT also starts accelerating but only up to 90 km/h, which was the ACC's initially set desired speed. Finally when the ALV starts braking, the ACC also activates the brakes and regulates the VUT's speed to 60 km/h.

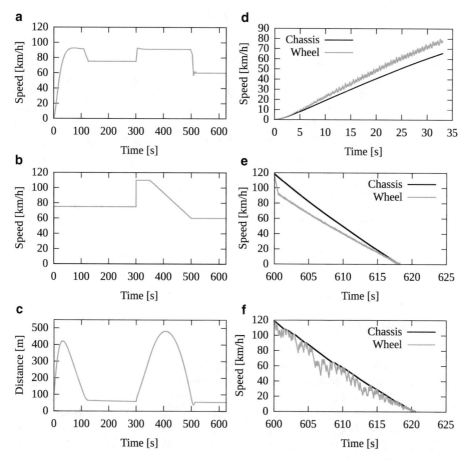

Fig. 1.12 Co-simulation results: standalone VP FMU scenario. (**a**) VUT speed. (**b**) TCS reaction. (**c**) ALV speed. (**d**) Correct ABS reaction. (**e**) Distance between vehicles. (**f**) Faulty ABS reaction

1.5.3.2 Test Scenario: ABS and TCS Road Test

To examine the operation of the ABS and TCS (both running on the VAP) in details, a similar co-simulation scenario has been assembled as follows:

- The road properties have been set to snowy conditions.
- After simulation start the VUT applies full throttle and tries to fully accelerate.
- Quite a while later the VUT applies full braking until complete standstill.

Since the snowy road causes the wheel to slip at acceleration, the TCS will be activated, that occasionally reduces engine torque and thus the wheel slip, as depicted in Fig. 1.12d. Finally, the ABS is triggered when applying full braking, that occasionally reduces the brake torque and thus the wheel slip, as depicted in Fig. 1.12e.

1.5.3.3 Test Scenario: ABS Fault Robustness

As a last scenario we wanted to evaluate the fault injection capabilities of PFMI. Thus, the wheel slip output of the vehicle model was corrupted by adding white noise to it, with an occurrence probability of 100 %, all via PFMI's fault injection pane. This false wheel slip signal was then fed to the VAP executing the ABS. As shown in Fig. 1.12f this resulted in an increased braking time (by around 2 s) until a complete standstill. After adding a noise filter for the input in the embedded ABS algorithm the braking behavior recovered to normal. This test intended to highlight the potentials of simulation-based fault injection, as a very efficient way to increase functional safety and add robustness to the embedded HW/SW system.

Finally, speed measurements have also been conducted for the HW/SW simulation separately. Here, the sc_start(fmu_steptime) function calls have been profiled, and the *real-time factor* (RTF) was measured, which is the wallclock time divided by the simulated time. While executing the driving scenarios, an average of RTF=22,4 has been achieved for the virtual platform simulation, that was executed on the host machine described in Sect. 1.5.1.

1.6 Related Work

Simulator coupling is an ongoing research topic. For instance, the authors of [6] and [22] propose guidelines to directly connect SystemC and *Simulink* simulations. Wawrzik et al. introduce a more generic framework to automatically generate domain- and tool-specific interconnects [37]. In [29] and [39] the authors present approaches and use-cases to achieve coupling between SystemC-based VPs, executing ADAS application prototypes, and 3D virtual driving simulators to evaluate them. However, all these works propose point-to-point solutions to connect simulation environments, which poses limitations on their flexibility, scalability, and integrability.

To address these drawbacks, simulation interoperability standards have been developed, and among them, FMI has become the de facto for loosely-coupled vehicular simulation systems. The dominance of FMI in the automotive industry is indicated by the numerous MBD tools that support FMU export and import. For example, the works presented in [1] and [21] strive for FMU model generation for two popular MBD tools: Simulink and *Jmodelica.org* [18]. According to [14], more than 60 MBD tools support FMU model export and FMI-based co-simulation possibility, which echoes the interest of the community.

Furthermore, the FMI co-simulation concept has been also thoroughly examined by the scientific community [5]. Broman et al. present comprehensive guidelines to ensure determinism, and propose control algorithms to achieve it [7]. In further works co-simulation possibilities of continuous-time vs. discrete event-based FMUs have been investigated [8, 23]. Such scientific contributions have continuously devoted and improved the FMI standard since its creation.

As mentioned previously, FMI defines a master/slave co-simulation structure. However, the FMI standard declares, but does not specify so-called *master algorithms* for orchestrating the co-simulation. Thus, numerous scientific contributions propose such algorithms and standalone simulation backplanes. For example, Bastian et al. propose the *EAS Master* as a standalone, multi-platform, simulator-coupling FMI backplane [4]. The composed prototype provides several algorithms for simulation coordination with fixed stepsize. Moreover, the authors have used OpenMP to execute the co-simulation in a parallel fashion, whenever possible. Nevertheless, the approach is hampered by the backplane's configuration possibilities, that limit the usability and flexibility of the module. In [2] and [3] Awais et al. propose to combine the advantages of FMI and another state-of-the-art co-simulation standard, the *High Level Architecture* (HLA) [13]. Here, they use an open-source implementation of an HLA simulation controller for FMI. Nonetheless, HLA poses limitations, since it supports network-distributed co-simulation only. In [27] *PyFMI* is presented, a basic, light-weight, cross-platform, Python-based FMI master module, also integrated in the MBD tool JModelica.org. But, since Python is an interpreted language, PyFMI might hamper simulation performance and memory utilization. As mentioned in Sect. 1.4, PFMI has been tailored to address the limitations of the aforementioned master modules and to ensure optimal simulation performance, while preserving interoperability and full-system analysis capabilities. On top of that, PFMI has been augmented with fault injection capabilities so to increase full-system robustness.

Finally, it is worth to mention, that several other co-simulation standards and tools have been developed such as [36] or [12]. Unfortunately most of them come with drawbacks, such as either being highly tool-specific, unnecessarily complex, or having a proprietary API, which makes their usage less appealing.

1.7   Summary and Outlook

This chapter presented two methods to integrate SystemC-based VPs into heterogeneous multi-domain vehicular simulation systems via the FMI standard:

- The tool-coupling VP FMU approach was demonstrated via the generation framework based on Synopsys' PA tool and evaluated in a practical vHIL co-simulation scenario with multiple FMUs using the Freescale Qorivva ECU VP.
- The standalone VP FMU concept was applied on the VAP within the top-down design flow for developing safety critical, reactive ADAS applications. The approach has also enabled to apply vHIL, integration and system testing of the algorithms in several driving scenarios involving multiple interconnected FMUs.

The chapter also presented the PFMI master for enhancing the performance of FMI-based multi-domain co-simulations. Furthermore, to ensure functional safety and SW robustness, the system-level fault injection capabilities of PFMI have been presented and evaluated with the VAP.

For future work, we plan to integrate standard-compliant, automated HW/SW testing facilities in PFMI. To enhance fault injection capabilities, we plan to introduce *deterministic* fault injection and more fault types so to increase the testing coverage. All these enhancements will lead to faster prototyping and stronger evaluation possibilities of the HW/SW subsystem of virtually developed vehicles.

References

1. Andersson, C., Åkesson, J., Führer, C., Gäfvert, M.: Import and export of functional mock-up units in JModelica.org. In: 8th International Modelica Conference (2011)
2. Awais, M.U., Palensky, P., Elsheikh, A., Widl, E., Matthias, S.: The high level architecture RTI as a master to the functional mock-up interface components. In: 2013 International Conference on Computing, Networking and Communications (ICNC), pp. 315–320, 28–31 January 2013. doi:10.1109/ICCNC.2013.6504102
3. Awais, M.U., Palensky, P., Mueller, W., Widl, E., Elsheikh, A.: Distributed hybrid simulation using the HLA and the functional mock-up interface. In: 39th Annual Conference of the IEEE Industrial Electronics Society (IECON 2013), pp. 7564–7569, 10–13 Novembar 2013. doi:10.1109/IECON.2013.6700393
4. Bastian, J., Clauß, C., Wolf, S., Schneider, P.: Master for co-simulation using FMI. In: 8th International Modelica Conference 2011, Dresden, pp. 115–120, 20–22 March 2011. doi:10.3384/ecp11063115
5. Blochwitz, T., Otter, M., Arnold, M., Bausch, C., Elmqvist, H., Junghanns, A., Mauss, J., Monteiro, M., Neidhold, T., Neumerkel, D., Olsson, H., Peetz, J. V., Wolf, S., Clauß, C.: The functional mockup interface for tool independent exchange of simulation models. In: 8th International Modelica Conference, pp. 105–114 (2011). doi:10.3384/ecp11063105
6. Boland, J.F., Thibeault, C., Zilic, Z.: Using matlab and Simulink in a SystemC verification environment. In: Proceedings of Design and Verification Conference (DVCon), San Jose, 14–16 February 2005
7. Broman, D., Brooks, C., Greenberg, L., Lee, E.A., Masin, M., Tripakis, S., Wetter, M.: Determinate composition of FMUs for co-simulation. In: Proceedings of the International Conference on Embedded Software (EMSOFT), 2013, pp. 1–12, 29 September 2013–4 October 2013. doi:10.1109/EMSOFT.2013.6658580
8. Broman, D., Greenberg, L., Lee, E.A., Masin, M., Tripakis, S., Wetter, M.: Requirements for hybrid cosimulation standards. In: Proceedings of the 18th International Conference on Hybrid Systems: Computation and Control (HSCC '15), pp. 179–188. ACM, New York (2015). doi:http://dx.doi.org/10.1145/2728606.272862910.1145/2728606.2728629
9. Broy, M.: Automotive software and systems engineering. In: Third ACM and IEEE International Conference on Formal Methods and Models for Co-Design, 2005. MEMOCODE '05, pp. 143–149, 11–14 July 2005. doi:10.1109/MEMCOD.2005.1487905
10. Bücs, R.L., Murillo, L.G., Korotcenko, E., Dugge, G., Leupers, R., Ascheid, G., Ropers, A., Wedler, M., Hoffmann, A.: Virtual hardware-in-the-loop co-simulation for multi-domain automotive systems via the functional mock-up interface. In: IEEE Forum on Specification and Design Languages (FDL), pp. 1–8, 14–16 September 2015. doi:10.1109/FDL.2015.7306355
11. Charette, R.N.: This car runs on code. In: IEEE Spectrum. http://spectrum.ieee.org/transportation/systems/this-car-runs-on-code. Accessed March 2014
12. COSIMATE official website. http://site.cosimate.com/. Accessed April 2015
13. Dahmann, J.S., Fujimoto, R.M., Weatherly, R.M.: The department of defense high level architecture. In: Proceedings of the 1997 Winter Simulation Conference, pp. 142–149, 7–10 December 1997. doi:10.1109/WSC.1997.640390
14. FMI standard official website. https://www.fmi-standard.org. Accessed April 2015

15. Freescale Semiconductors: Freescale Qorivva MPC5643L data sheet rev. 9. www.freescale. com/webapp/sps/site/prod_summary.jsp?code=MPC564xL. Accessed June 2013
16. Georgakos, G., Schlichtmann, U., Schneider, R.: Reliability challenges for electric vehicles: from devices to architecture and systems software. In: 2013 50th ACM/EDAC/IEEE Design Automation Conference (DAC), pp. 1–9, 29 May 2013–7 June 2013. doi:10.1145/2463209.2488855
17. ISO 26262 "Road vehicles – functional safety" standard. www.iso.org/iso/catalogue_detail? csnumber=43464. Accessed April 2015
18. JModelica.org official website. http://www.jmodelica.org/. Accessed April 2015
19. Jones, D., Topham, N.: High speed CPU simulation using LTU dynamic binary translation. In: Seznec, A., Emer, J., O'Boyle, M., Martonosi, M., Ungerer, T. (eds.) Proceedings of the 4th International Conference on High Performance Embedded Architectures and Compilers (HiPEAC '09). Springer, Berlin/Heidelberg, pp. 50–64. doi:http://dx.doi.org/10.1007/978-3-540-92990-1_6 10.1007/978-3-540-92990-1_6
20. Kooli, M., Di Natale, G.: A survey on simulation-based fault injection tools for complex systems. In: 9th IEEE International Conference on Design & Technology of Integrated Systems in Nanoscale Era (DTIS), pp. 1–6, 6–8 May 2014. doi:10.1109/DTIS.2014.6850649
21. Lang, J., Rünger G., Stöcker P.: Dynamische Simulationskopplung von Simulink-Modellen durch einen Functional-Mock-up-Interface-Exportfilter. Student research project, 2013 Faculty of Computer Science, Technical University of Chemnitz, Germany. www.tu-chemnitz.de/ informatik/service/ib/pdf/CSR-13-05.pdf. Accessed April 2015
22. Mendoza, F., Kollner, C., Becker, J., Muller-Glaser, K.D.: An automated approach to SystemC/Simulink co-simulation. In: 22nd IEEE International Symposium on Rapid System Prototyping (RSP), 2011, pp. 135–141, 24–27 May 2011. doi:10.1109/RSP.2011.5929987
23. Muller, W., Widl, E.: Linking FMI-based components with discrete event systems. In: 2013 IEEE International Systems Conference (SysCon), pp. 676–680, 15–18 April 2013. doi:10.1109/SysCon.2013.6549955
24. OpenModelica official website. https://openmodelica.org/. Accessed April 2015
25. OpenMP specification. http://openmp.org/wp/openmp-specifications/. Accessed April 2015
26. Pretschner, A., Broy, M., Kruger, I.H., Stauner, T.: Software engineering for automotive systems: a roadmap. In: Future of Software Engineering, 2007. FOSE '07, pp. 55–71, 23–25 May 2007. doi:10.1109/FOSE.2007.22
27. PyFMI official website. https://pypi.python.org/pypi/PyFMI. Accessed April 2015
28. SafeCar website - National Highway Traffic Safety Administration, United States Department of Transportation. www.safercar.gov. Accessed April 2015
29. Schneider, S.A., Frimberger, J.: Significant reduction of validation efforts for dynamic light function with FMI for multi-domain integration and test platforms. In: 10th International Modelica Conference, Lund, 10–12 March 2014. doi:10.3384/ecp14096395
30. Schutter, T.D.: Better Software. Faster!: Best Practices in Virtual Prototyping. Synopsys Press, Mountain View (2014)
31. Simulink Simscape multi-domain physical system library: uk.mathworks.com/products/ simscape/index.html. Accessed April 2015
32. Simulink Stateflow vehicle model: mathworks.com/help/stateflow/gs/how-stateflow-software-works-with-simulink-software.html. Accessed April 2015
33. Synopsys' Platform Architect tool: www.synopsys.com/Prototyping/ArchitectureDesign/ pages/platform-architect.aspx. Accessed April 2015
34. Synopsys: Virtualizer third-party integration manual for Synopsys Saber, Cadence AMS Designer, MathWorks Simulink and Vector CANoe. Accessed March 2014
35. SystemC standard: Accellera official website. http://www.accellera.org/downloads/standards/ systemc. Accessed April 2015
36. The MathWorks Inc.: Developing S-Functions manual, version 8.4 (release 2014b). http://jp. mathworks.com/help/pdf_doc/simulink/sfunctions.pdf. Accessed October 2014

28 R.L. Bücs et al.

37. Wawrzik, F., Chipman, W., Molina, J.M., Grimm, C.: Modeling and simulation of cyber-physical systems with SICYPHOS. In: 10th International Conference on Design & Technology of Integrated Systems in Nanoscale Era (DTIS), pp. 1–6, 21–23 April 2015. doi:10.1109/DTIS.2015.7127375
38. Weinstock, J.H., Schumacher, C., Leupers, R., Ascheid, G., Tosoratto, L.: Time-decoupled parallel SystemC simulation. In: Design, Automation and Test in Europe Conference and Exhibition (DATE), 2014, pp. 1–4, 24–28 March 2014. doi:10.7873/DATE.2014.204
39. Wehner, P., Ferger, M., Gohringer, D., Hubner, M.: Rapid prototyping of a portable HW/SW co-design on the virtual Zynq platform using SystemC. In: IEEE 26th International SOC Conference (SOCC), 2013, pp. 296–300, 4–6 September 2013. doi:10.1109/SOCC.2013.6749704

Chapter 2
Standard Compliant Co-simulation Models for Verification of Automotive Embedded Systems

Martin Krammer, Helmut Martin, Zoran Radmilovic, Simon Erker, and Michael Karner

2.1 Introduction

Cooperative simulation, or co-simulation, has become a common method to support the development of automotive systems. The integration of different modelling languages, tools and solvers into one common co-simulation enables new possibilities for design and verification of complex systems. Efforts to standardize the exchange of simulation models and enable integration in co-simulation scenarios were undertaken by the ITEA2 MODELISAR project. One of its main goals was the development of the functional mock-up interface[1] (FMI) [5, 14]. The FMI is an open standard which defines an interface supporting model exchange between simulation tools and interconnection of simulation tools and environments. The second version of the FMI standard was released in 2014 [6, 15].

SystemC[2] [19] is a C++ based library for modelling and simulation purposes. It is intended for the development of complex electric and electronic systems. SystemC targets high abstraction level modelling for fast simulation [2]. It provides sets of macros and functions, and supports paradigms like synchronization, parallelisms, as well as inter-process-communications. Its simulation engine is included in the library, and is built into an executable during model compilation. While

---

[1]http://www.fmi-standard.org.

[2]http://www.accellera.org.

M. Krammer (✉) • H. Martin • Z. Radmilovic • S. Erker • M. Karner
Virtual Vehicle, Inffeldgasse 21a, 8010 Graz, Austria
e-mail: martin.krammer@v2c2.at; helmut.martin@v2c2.at; zoran.radmilovic@v2c2.at; simon.erker@v2c2.at; michael.karner@v2c2.at

© Springer International Publishing Switzerland 2016
R. Drechsler, R. Wille (eds.), *Languages, Design Methods, and Tools for Electronic System Design*, Lecture Notes in Electrical Engineering 385, DOI 10.1007/978-3-319-31723-6_2

SystemC is capable of modelling and simulating digital systems, its SystemC-AMS[3] extension expands these concepts to the analog and mixed signal domain. Both, SystemC and SystemC-AMS libraries, provide a certain degree of protection of intellectual property, when optimized and compiled models are exchanged.

In this work, we present a tool-independent method on how to integrate electric and electronic system models together with their corresponding simulation engines into single functional mock-up units (FMU) implementing the FMI. Aforementioned models are built using SystemC and SystemC-AMS. By doing so, SystemC becomes available to a broad range of applications on system level in a standardized manner. The resulting FMUs are highly transportable and may easily be integrated within larger and more complex co-simulation scenarios for fast and convenient information exchange and system verification.

This paper is structured as follows. Section 2.2 recapitulates related work. Section 2.3 characterizes relevant frame conditions and requirements. Section 2.4 introduces necessary steps on how to process models for FMU integration. Section 2.5 highlights the application of the proposed method with an automotive battery system use case. Section 2.6 summarizes the results and concludes this paper.

2.2   Related Work

Since the release of the FMI standard version 1.0 in 2010 and version 2.0 in 2014, efforts have been spent in order to implement and test the functional mock-up interface, in order to build new workflows for simulation and verification of systems under development. This section presents related work in the area of FMI, FMU generation, as well as parsing and usage in simulation scenarios.

Corbier et al. [10] introduces the FMI and argues about the necessity to share models for model/software/hardware-in-the-loop testing activities. As part of that a methodology for gradual integration and progressive validation is proposed. It also emphasises the need for conversion of existing models into the FMI standard.

Chen et al. [9] discusses technical issues and implementation of a generic interface to support the import of functional mock-up units into a simulator. For this import, the FMI calling sequence of interface functions from the standard are used.

Noll and Blochwitz [27] describes the implementation of FMI in SimulationX. It presents code generation out of a simulation model for FMUs for model exchange and co-simulation. A code export step generates the necessary C-code for model exchange. For co-simulation, a solver is included in the resulting dynamic link library (DLL). The tool coupling using SimulationX is accomplished by using a wrapper.

---

[3]http://www.systemc-ams.org.

The need for co-simulation in connection with the design of cyber-physical systems is highlighted in [32]. It follows the idea, that coded solvers in FMUs have some limitations regarding analysis or optimization. Therefore the authors strive for explicitly modelled ordinary differential equation solvers and claim a significant performance gain.

In [7] a verification environment using Simulink and SystemC is introduced. It relies on s-functions to create a wrapper in order to combine SystemC modules with Simulink.

In [30] the generation of FMUs from software specifications for cyber-physical systems is outlined. This approach fulfills the need for software simulation models. A UML based software specification is automatically translated into a FMU, maintaining its original intended semantics. This step is done using C-code, which is included within the FMU.

In [25] a high level approach for integration and management of simulation models for cyber-physical systems is shown.

Elsheikh et al. [13] presents an integration strategy for rapid prototyping for Modelica models into the FMI standard, and highlights a high level approach for integration of cyber-physical systems.

SystemC and SystemC-AMS are used in a variety of simulation platforms, where different wrappers or adapters provide data exchange services. Examples thereof are given in [1, 20]. Regarding the use of SystemC or SystemC-AMS in the context of the FMI standard, no relevant publications are available to date, describing a unified process for integration. Thus, the outlined approach of integrating SystemC/SystemC-AMS models together with the functional mock-up interface into a FMU is considered as a novel contribution to the field of applied co-simulation.

2.3 Requirements on Modelling and Co-simulation Data Exchange

In this section, general framework requirements and specifics are captured. This affects the SystemC and SystemC-AMS languages and libraries, the FMI standard, as well as co-simulation specific aspects. Usually, the compilation of SystemC or SystemC-AMS models leads to a platform specific executable, containing all models as well as necessary schedulers and solvers. This property seems suitable for application of SystemC's modelling and simulation concepts in context of the FMI. In order to comply to the FMI standard for co-simulation, a dynamic link library implementing the FMI needs to be compiled and assembled instead of an executable. The targeted FMI application scenario can be found in [14, 15] and is shown in Fig. 2.1 for one single FMU. One co-simulation master is expected to coordinate the co-simulation by utilizing the FMI for communication to the FMU. The FMU

Fig. 2.1 Co-simulation with
generated code on a single
computer [15]

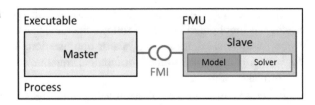

includes the entire model and the corresponding scheduler or solver. SystemC and
SystemC-AMS based approaches differ from co-simulation with tool coupling, as
no separate tool is required for simulation execution.

A state machine for the calling sequence of the FMI standard co-simulation
interface C-functions is available [14, p. 31]. For a basic repetition of simulation steps, the following principles can be distilled. First, a FMU is subject
to instantiation and initialization. Then, input values, parameters and states are
set using corresponding fmiSet[...]() group of functions. This is followed
by a simulation execution phase called using fmiDoStep() function. If that
step completes, the fmiGet[...]() group of functions is used to retrieve the
results for external communication. In scope of this work the previously mentioned
functions need to be implemented in SystemC and made available to the FMI based
on a C-code interface. Since SystemC supports the concept of time, all C-interface
function calls must be synchronized during co-simulation.

One further main criteria refers to the SystemC and SystemC-AMS libraries.
Both libraries are available license free. SystemC is standardized under
IEEE 1666 [19], whereas SystemC-AMS is documented in the Language Reference
Manual version 2.0. Thus, it makes sense to strive for a solution which builds on
these standard documents and does not cause any changes to the corresponding
implementation libraries.

Under normal circumstances, SystemC and SystemC-AMS module simulations
are performed in a single run, using the time domain simulation analysis mode. This
means that the method sc_start() runs the initialization phase and subsequently
the scheduler through to completion [11]. However, sc_start() may be called
repeatedly with a time argument, where each simulation run starts where the
previous run left off. For co-simulation, where data exchange and synchronization
happens on discrete points in time, this function is vital to control simulation within
the FMU. Memory management is rarely an issue in standard SystemC/SystemC-
AMS models, due to their single elaboration phase and comparably short run times.
In encapsulated FMUs, memory management can be crucial as the FMI standard
suggests that FMUs have to free any allocated resources by themselves. The
standard therefore defines instantiation and termination functions, therefore proper
construction and destruction of all SystemC/SystemC-AMS models contained in
FMUs is desirable to avoid any memory leaks.

The SystemC simulation kernel supports the concept of a delta cycle. One delta
cycle consists of an evaluation phase followed by an update phase. This separation
ensures deterministic behavior [11], as opposed to e.g. the use of events. Events

trigger process executions, but their execution order within one single evaluation phase is non-deterministic. The concept of a delta cycle is even more important when moving from simulation level to co-simulation level, where external signals are connected to the model. New values written to e.g. signals become visible after the following delta cycle. Execution of one delta cycle does not consume simulated time.

SystemC and SystemC-AMS feature four different models of computation (MoC). According to [34], a MoC is defined by three properties. First, the model of time employed. Second, the supported methods of communication between concurrent processes. And third, the rules for process activation. SystemC features a kernel including a non-preemptive scheduler [19], which operates discrete event (DE) based for modelling concurrency. This MoC is used for modelling and simulation of digital software and hardware systems. SystemC-AMS features three different MoC, which may be used depending on the actual application. Namely these are timed dataflow (TDF), linear signal flow (LSF) and electrical linear networks (ELN). TDF operates on samples which are processed at a given rate, with a specified delay, at a given time step interval. TDF and DE MoC are synchronized using specified converter ports. LSF is primarily used for signal processing or control applications and features a broad range of predefined elements within the SystemC-AMS library. Converter modules for the conversion to and from the TDF and DE MoC exist. ELN permits the description of arbitrary linear networks and features an element library as well. Converter modules for the conversion to and from the TDF and DE MoC exist. Our goal is to support all four MoC in context of simulation through the FMI.

2.4   Model Integration Method

In order to integrate and execute SystemC and SystemC-AMS simulation models in context of an FMU, we propose a structured method. The necessary steps are illustrated in Fig. 2.2, indicated by the dashed box. They are described as follows.

(A) Modelling and simulation of a single component model.
(B) Model interface identification for coupling to the co-simulation environment.
(C) Wrapper class specification for controlling the model interface.
(D) C-interface specification for FMI integration.
(E) FMI integration using a predefined software developer kit.
(F) FMU compilation and assembly together with (architectural) model description.
(G) Integration of FMU to co-simulation master for simulation based system level verification.

Subsequently, each step is explained in detail.

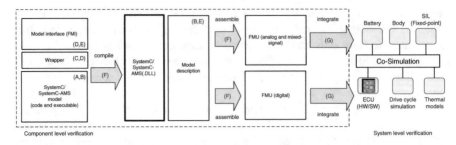

Fig. 2.2 The proposed process for the integration of executable simulation models into FMUs for co-simulation

2.4.1 Modelling and Simulation

To embed a SystemC or SystemC-AMS simulation model in a FMU, the model has to be set up and tested against its specifications first. Standalone executables may be executed, traced, and debugged using additional tools like an integrated development environment. However, a proprietary co-simulation master usually does not offer such sophisticated debug possibilities for compiled and assembled FMUs. To instantiate and test the model under development, a test bed is typically used. It may consist of stimuli generators, reference models or watchdogs [16]. The *SystemC Verification Library* (SCV) [35] is also available for this purpose.

2.4.2 Model Interface Identification

In a second step, the interface of the simulation model, which is later exposed through the FMI, shall be determined. This includes the definition of input and output quantities, or states, as well as internal timing (accuracy and precision required by the simulation model) and external timing (simulation step size for data exchange considerations. If a system level design is available, the model interface identification can be accomplished using these specifications.

2.4.3 Wrapper Class Specification

Typically, `sc_main()` is used to indicate the top-level module and to subsequently construct the entire module hierarchy through instantiation. The latter happens during the so called elaboration phase, right before the execution of `sc_start()`. For co-simulations, the breakdown of time into time steps is achieved by calling `sc_start()` multiple times. Thus it is necessary to keep the entire module hierarchy and its states persistently in memory, as seamless data exchange between simulation steps must be ensured. This can be achieved by using a C++ wrapper class defining a constructor and destructor managing the top-level SystemC simulation model in memory, until the entire co-simulation is finished. Listing 2.1 shows an example constructor for a SystemC wrapper class. Additionally, the wrapper class has sc_signal primitive channels attached. These are used for realization of the interface identified in step (B).

```
BatteryControllerWrapper :: BatteryControllerWrapper () {
controller  = new  BatteryController("batteryController");

signalVoltageIn  =  new  sc_signal <double >;
signalSocOut  =  new  sc_signal <double >;
controller ->in_voltage (* signalVoltageIn );
controller ->out_soc (* signalSocOut );
}
```

Listing 2.1 Wrapper class constructor.

2.4.4 C-Interface Specification

The main idea here is to have a set of functions, which can be used to initialize, control, and finally shut down the C++-based SystemC/SystemC-AMS simulation model within a C-based FMU. To bridge the gap between the C-language FMI and the C++-language based simulation models, special linker instructions are required when compiling the file for the FMU. The compiler keyword used for this purpose is extern "C". This C++ standard feature is a linkage-specification every compiler is required to fulfill. It exposes the enclosed C++ functions to the FMI.

For SystemC/SystemC-AMS, this means that signals (e.g. sc_signal) or ports (e.g. sc_in/sc_out) may be used for variable modifications when the scheduler or simulation engine is not running. However, the scheduler requires the execution of one delta cycle to adopt a value which is written to a signal, otherwise the previous value remains assigned. This is achieved by using the SC_ZERO_TIME macro. It updates the signal's value while it does not advance simulation time.

For SystemC-AMS, the different MoC are combined advantageously using converters, in order to get and set input and parameter values as desired. To couple e.g. an electric current, the sca_eln::sca_tdf_isink and sca_eln::sca_tdf::sca_r primitive modules from the ELN MoC library may be used to read or write electric current values using the TDF MoC, respectively. A code example for setting a value to a TDF MoC module can be seen in Listing 2.2. Again, for execution of one delta cycle, the SC_ZERO_TIME macro is used.

```
extern "C" void setBatteryCurrent(double current) {
wrapper->batteryModule->cellGenerator->setCurrent(current);
sc_start(SC_ZERO_TIME);
}
```

Listing 2.2 Assignment of a value to a TDF module method

For FMI integration, a minimum of six different kinds of functions are required. `startInterface()` is required as an entry point to the SystemC/SystemC-AMS model, when simulation is initialized. Called once per FMU instantiation, this function calls `sc_main()` for the first time. One delta cycle is increased by calling `sc_start()` with the `SC_ZERO_TIME` argument. This triggers the construction of the simulation model in memory via the previously introduced wrapper class from step (C). This ensures that the simulation model is completely hierarchically constructed in memory and ready for simulation, without any simulated time passing by yet. `shutDownInterface()` is used to destruct all impressed models and free the occupied memory. `setValueXXX()` and `getValueXXX()` functions are used for each coupled variable to pass values through the FMI directly to the SystemC/SystemC-AMS model. The `doSim()` function is used to trigger the simulation start. It basically calls `sc_start()` with a SystemC time format parameter. If `sc_start()` is called using a fixed time interval, this time interval represents the step size of the FMU. In order to dynamically pass a required time interval setting to the simulation model, the `setTimeStep()` function is used. This realizes an adaptive step size co-simulation. The time step value is stored within the wrapper class. The call to `sc_start()` is modified accordingly, as seen in Listing 2.3.

```
sc_core::sc_start(wrapper->timeStep, sc_core::SC_SEC);
```

Listing 2.3 Dynamic simulation step size assignment.

2.4.5 FMI Integration

For integration of the FMI, the FMU SDK (software development kit) is used as a basis [33]. It provides functions and macros which are included next to the SystemC/SystemC-AMS files. The main header file establishes a logical connection between the software code and the descriptive XML file that is required for each FMU. This `modelDescription.xml` file contains major information about the FMUs architecture. This also includes gathered information from step (B).

2.4.6   FMU Compilation and Assembly

Finally, the FMU parts are now compiled and assembled. According to [14, p. 38], a FMU is referred to as a zip file with a predefined structure. The .dll file is placed in the binaries folder for the corresponding platform. The modelDescription.xml file is placed in the top level (root) folder. The SystemC/SystemC-AMS source files are placed in the sources folder optionally. Corresponding model documentation or associated requirements (see [21] for details) may be placed in the documentation folder. Initialization values, like sets of parameters, may be placed in the resources folder. After creating a zipped file from these contents, the FMU is ready for distribution and instantiation.

2.4.7   FMU Integration for Co-simulation

The FMI for co-simulation standard defines a master software component which is responsible for data exchange between subsystems. In this work, the independent co-simulation framework (ICOS) from *Virtual Vehicle Research Center* [18] is used. Typically, it is applied to solve multidisciplinary challenges, primarily in the field of automotive engineering. Use cases include integrated safety simulation, electrical system simulation, battery simulation, thermal simulation, mechanical simulation, and vehicle dynamics simulation. It features an application design which separates the co-simulation framework and its coupling algorithms from the simulation tools that are part of the co-simulation environment. Hence the co-simulation framework is independent from the simulation tools it integrates [31]. The framework relies on the exchange of discrete time signal information, and provides several different algorithms e.g. for interpolation, extrapolation, and error correction methods [3] of these signals. The exchange of discrete time signals from one co-simulation component model to another is called coupling. The independent co-simulation framework implements the FMI standard and allows instantiation and co-simulation of FMUs. Alternatively, and to ensure FMI standard compliance, a *FMU checker* executable is provided with the standard. It instantiates an FMU and performs basic operations on it automatically.

2.5   Automotive Battery System Use Case

In order to demonstrate the functionality of the resulting FMUs described in Sect. 2.4, an automotive battery system use case was selected. The battery system introduced here is intended for a hybrid electric vehicle (HEV), where the battery powers an electric motor next to a combustion engine, as main components of the power train. Several strategies are known to charge and discharge an automotive battery propelling a vehicle. Information about driver behavior, routing, road profile, etc. have a strong influence on the behavior of the battery system.

Such a battery system usually consists of two main components, namely the battery pack as energy storage facility and a corresponding battery controller. The targeted battery pack consists of four battery modules, where each of them integrates 12 cells with a nominal capacity of 24Ah each. The targeted controller monitors operational condition and health of each of the modules. The controller also communicates with the car's hybrid control unit (HCU) to ensure stable vehicle operation. For the battery module, we construct a SystemC-AMS based FMU utilizing the ELN and TDF MoC. For the battery controller, we implement a SystemC/SystemC-AMS based FMU utilizing the DE and LSF MoC. Both FMUs are integrated into one common co-simulation scenario.

2.5.1 Battery Module FMU

The success of HEVs strongly depends on the development of battery technology. For the development of battery based systems it is essential to know the characteristics and the behavior of the battery. Especially in the field of functional safety, the knowledge of these features is key to control potential hazards. There is a wide range of different existing modes available for simulating the performance of lithium-ion battery cells. Most common approaches are electro-chemical models [12, 26, 29] and equivalent circuit models (ECM) [4, 8, 17, 24].

Electrochemical models describe the internal dynamics of a cell leading to complex partial differential equations with a large number of unknown parameters. Since these models are computationally expensive and therefore time consuming, they are not suitable for system-level modeling. ECMs on the other hand are computationally more efficient and are therefore better suitable for system-level modeling. Such models are also commonly used in embedded battery management systems (BMS) to estimate the state of charge (SoC) and predict the performance of physical batteries. An ECM is composed of primitive electrical components like e.g. resistors, capacities and voltage sources to simulate a cell's terminal voltage response to a desired current flow. It is capable to accurately describe the static as well as the dynamic behavior of a cell under various operating conditions. In [36] an ECM model of a battery cell intended for portable devices, written in SystemC-AMS, is shown.

For this use case, we model an ECM according to [24, 36]. The corresponding circuit diagram is shown in Fig. 2.3. Such a two-RC block model (a) is a common choice for lithium-ion cells [17]. These two RC blocks characterize short-term and long-term dynamic voltage response of the cell, respectively, which arise from diffusion phenomena in the cell. I_{bat} in (b) represents an identified input current to the FMU according to Sect. 2.4. The controlled voltage source V_{OC} reproduces the open circuit voltage (OCV), which represents an output of the FMU according to Sect. 2.4. The serial resistor R_0 describes the internal resistance of the cell comprised

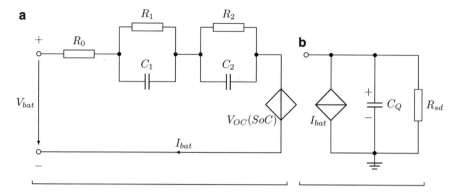

Fig. 2.3 Schematic diagram of the used cell equivalent circuit model: (**a**) voltage-current characteristics, (**b**) energy balance circuit

of ohmic and charge transfer resistances and is connected in series with the two RC branches. In general, all parameters of the model depend on several quantities like state of charge (SoC), temperature, cell age as well as current direction and rate.

Mathematically the electrical behavior of an ECM with two RC branches can be expressed by Eqs. 2.1 and 2.2.

$$V_{bat} = V_{OC} + V_1 + V_2 + R_0 I_{bat} \tag{2.1}$$

$$\begin{bmatrix} \dot{V}_1 \\ \dot{V}_2 \end{bmatrix} = \begin{bmatrix} \frac{1}{R_1 C_1} & 0 \\ 0 & \frac{1}{R_2 C_2} \end{bmatrix} \begin{bmatrix} V_1 \\ V_2 \end{bmatrix} + \begin{bmatrix} \frac{1}{C_1} \\ \frac{1}{C_2} \end{bmatrix} I_{bat} \tag{2.2}$$

with $V_{1,2}$ and $\dot{V}_{1,2}$ the voltages across the RC branches and their time derivatives, respectively.

To parametrize an ECM, measurements on physical batteries are often performed to create large multidimensional look-up tables for the various parameters to cover all their dependencies. The lithium-ion battery model is based on data of a LiFePO$_4$ (LFP) battery [24]. To limit complexity for this use case we propose a simplified model, where the OCV depends on the SoC, $V_{OC} = f(SoC)$, and all other parameters are considered constant using averaged values from the data of [24]. The used $SoC - V_{OC}$ relationship is based on [23] and was adapted for usage out of its ordinary range to support simulation of e.g. overcharge scenarios. Figure 2.4 shows the $SoC - V_{OC}$ relationship. However, the model can straightforwardly be extended to a more advanced model by including additional elements representing terms for e.g. thermal behavior or aging effects. Thermal models may also be attached through co-simulation later on.

It is noteworthy that the ECM model can be adopted for other battery chemistries by simply adjusting the set of parameters used in the model. In context of a FMU, these could be placed within the FMU's resources folder in e.g. XML file format for model configuration during instantiation.

Fig. 2.4 Battery open circuit
voltage V_{OC} as a function
of *SoC*

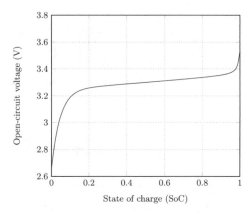

From a black box view, the proposed battery module FMU shall support the
following functionality.

- One module shall consist of 12 cells in series connection, each modelled as
  described in this section.
- Current demand is an input to the battery module.
- Voltage is an output of the battery module.
- The battery module's SoC shall be set from external prior to simulation (without
  recompilation of the FMU).

A battery module with these properties is modelled in SystemC-AMS as
follows. The ECM of a battery cell shown in Fig. 2.3 representing Eqs. 2.1, 2.2
is implemented using primitives of the ELN MoC of SystemC-AMS. The TDF
MoC is used to interface the ECM, e.g. to control the voltage source V_{OC} the class
`sca_tdf_vsource` primitive is used. The cell is instantiated 12 times, and these
instances are connected in series to model a single battery module. The battery
module is subject to integration into a FMU according to Sect. 2.4.

2.5.2 Battery Controller FMU

The electrochemical processes inside a battery are considered complex and the
lithium-ion cell operating window is narrowed down by voltage and temperature
restrictions. A battery controller monitors cells and modules to keep the battery
pack and the entire vehicle functional within a safe state. Monitoring measures
for batteries typically capture the state of charge (SoC), cell voltages as well as
the temperature. The determination of the SoC is very complex, since it cannot
be measured directly. Many vehicle applications require an exact knowledge of
the SoC of the battery. The most obvious include the calculation of the vehicle's
remaining driving range. More sophisticated applications include communications
to the HCU, e.g. recuperation functions, influence on driving modes, trip routing, or

comfort functions. Several solutions to the problem of accurately estimating the SoC have been proposed in literature [28]. The most common method for calculating the SoC is coulomb counting, which is based on measuring battery current. With the knowledge of an initial SoC_0, the remaining capacity in a cell can be calculated by integrating the current that is entering (charging) or leaving (discharging) the cell over time:

$$SoC = SoC_0 + \frac{1}{C_Q} \int_{t_0}^{t} \eta I_{bat}(\tau) \cdot d\tau \qquad (2.3)$$

Here C_Q is the rated capacity (the energy capacity of the battery under normal condition), I_{bat} is the battery current and η is a factor that accounts for loss reactions in the cells (we assume $\eta = 1$). Coulomb counting is straightforward to implement and able to determine the SoC under load, which makes it suitable for on-board applications. However, it requires the initial SoC of the battery. To get SoC_0 the cell voltage under no-load is measured and from this the SoC can be determined from the $SoC - V_{OC}$ relationship.

We propose a battery controller FMU, implementing four main functionalities:

- Sampling of the battery voltage.
- Sampling of the battery current.
- Calculation of the SoC based on $SoC - V_{OC}$ relationship and look-up table.
- Calculation of the SoC based on coulomb counting using current integration.

The battery controller is realized as a SystemC/SystemC-AMS module. It uses one thread for periodical voltage sampling and one for look-up table operations (based on the relationship depicted in Fig. 2.4). It utilizes the integrator primitive from the LSF MoC of SystemC-AMS for current flow calculation according to Eq. 2.3. In the end, the module is compiled and assembled as an FMU according to Sect. 2.4.

2.5.3 Co-simulation Integration

The resulting FMUs pass the *FMU checker* test and are integrated into the independent co-simulation framework as described in Sect. 2.4.7. First, its boundary condition server (BCS) is used to test the resulting single FMUs. Second, the FMUs are integrated into a co-simulation scenario. This includes a mild hybrid vehicle together with a drive cycle modelled in CarMaker, and the hybrid controls modelled in Matlab Simulink.

The integration is accomplished using a computer aided software engineering (CASE) tool, namely Enterprise Architect.[4] It uses a co-simulation framework

---

[4]http://www.sparxsystems.com.

specific extension [22] based on UML and SysML languages to integrate the required simulation models, assign configuration information to them, and generate a co-simulation configuration file out of the underlying model repository. For model integration, the model interface information described in Sect. 2.4 (B) is used. The simulation models and their couplings are represented in a co-simulation architecture diagram (cad). This is shown in Fig. 2.5. The couplings between models are defined using relationships at their interfaces. In order to ensure the correct transformation from the SysML model to the co-simulation configuration file, a model analyzer checks for presence of modelling errors. The co-simulation itself is executed through the independent co-simulation framework, which consumes the generated configuration file.

2.5.4   Discussion of Results and Observations

The following Table 2.1 provides an overview of different simulation scenarios and their achieved performance within the independent co-simulation framework as described in Sect. 2.4.7. The simulated time is denoted as t_s, whereas t_e refers to the time needed for simulation execution on a standard laptop device.

Figure 2.6 shows the output of the controller's estimation of the SoC. In this scenario, the module is discharged with 20 A current pulses with a pulse period of $T = 1100\,s$ and a duty factor of $\tau/T = 1/11$. State of charge estimation with a simple OCV-SoC lookup table does not make sense under load conditions, as during times $T - \tau$. Allowing a sufficiently large relaxation time after a discharge pulse we can compare the two methods and yield similar results for the SoC. Next, the controller and battery module FMUs are coupled to the vehicle model and its hybrid controls. Figure 2.7 shows an excerpt of the electric motor current demand and battery module state-of-charge as observed during a drive cycle.

From a qualitative point of view, the battery simulation results correspond to the results described in [23]. Quantitatively, the generated battery module FMU reproduces the simulation results shown in [36], validating against Li-Po cells from [9]. For this case no relevant increase of simulation time caused by the co-simulation framework was measured. The step size considered for co-simulation was 1 s.

The ascertained overall efforts for FMU integration are considered justifiable, once the co-simulation interface has been defined. However, the following issues should be observed when following the proposed process. By encapsulating simulation models into FMUs, an additional layer of time synchronization is introduced. FMU-internal SystemC/SystemC-AMS module step sizes may take very small values and account for precise calculations. In contrast to that, a very small external co-simulation step size causes increased coupling-related communications, produces vast amounts of data, and therefore slows down simulation performance. From this it follows that the internal and external step sizes used may not diverge by higher orders of magnitude.

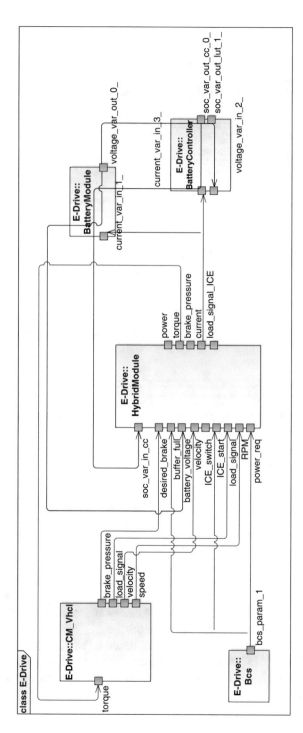

Fig. 2.5 Model integration

Table 2.1 Simulation time evaluation

Scenario	t_s [s]	t_e [s]
Battery controller w/ discharge pulses (BCS)	8000	9
Battery controller w/ drive cycle current (BCS)	200	<1
Battery module w/ discharge pulses (BCS)	8000	2
Battery module w/ drive cycle current (BCS)	200	<1
Battery module, controller w/ discharge pulses (BCS)	10, 000	15
Battery module, controller w/ discharge pulses (BCS)	20, 000	28
Mild hybrid vehicle drive cycle	200	17

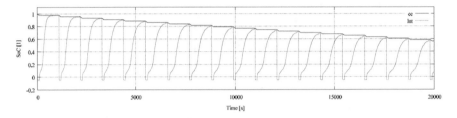

Fig. 2.6 Estimation of the state-of-charge using voltage based look-up table (lut) and coulomb counting with current integration (cc) approaches. For this simulation, a 20 A pulse discharge test was conducted

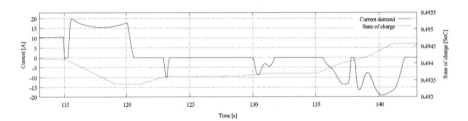

Fig. 2.7 The electric motor current demand (*solid line*) and battery module state-of-charge (*dashed line*) of a hybrid electric vehicle as observed during a drive cycle. The co-simulation scenario includes a vehicle power train model and a hybrid control unit model coupled to the two proposed FMUs

To ensure stable loading, execution, and unloading processes of FMUs at the FMI master, the use of pointers and dynamic memory allocation when constructing SystemC/SystemC-AMS modules is indispensable.

The FMI standard defines a resource folder inside a FMU, which may be used for e.g. different sets of ECM parameter settings. This is ideal for model exchange scenarios where models are kept separately from their associated parameters and configurations.

The creation of the required FMU XML file and synchronization to the model code causes additional efforts due to variable numbering and name assignments, especially if models are modified. Additional automation could help to improve these deficiencies, e.g. use of a model based software development approach.

2.6 Conclusion

In this paper a structured method for the integration of SystemC/SystemC-AMS simulation models to the FMI standard is introduced. The presented method does not require any changes to the standardized SystemC or SystemC-AMS libraries. The method eases the integration of existing system level simulation models into larger and more complex simulation scenarios, which are used for information exchange and verification on system level. A two-part battery system use case from the automotive domain is presented, which exploits these MoC for simulation. The resulting FMUs created with the described method are highly transportable and configurable. These properties make them suitable for verification and information exchange processes within the automotive domain.

Acknowledgements This research work has been funded by the European Commission within the EMC^2 project under the ARTEMIS Joint Undertaking under grant agreement no. 621429. The authors acknowledge the financial support of the COMET K2—Competence Centres for Excellent Technologies Programme of the Austrian Federal Ministry for Transport, Innovation and Technology (BMVIT), the Austrian Federal Ministry of Science, Research and Economy (BMWFW), the Austrian Research Promotion Agency (FFG), the Province of Styria and the Styrian Business Promotion Agency (SFG).

References

1. Armengaud, E., Karner, M., Steger, C., Weiß, R., Pistauer, M., Pfister, F.: A cross domain co-simulation platform for the efficient analysis of mechatronic systems. In: SAE World Conference (SAE Technical Paper 2010-01-0239), pp. 1–14 (2010). doi:10.4271/2010-01-0239
2. Barnasconi, M.: SystemC AMS extensions: solving the need for speed. In: Design Automation Conference (2010)
3. Benedikt, M., Watzenig, D., Zehetner, J., Hofer, A.: NEPCE - a nearly energy preserving coupling element for weak-coupled problems and co-simulation. In: IV International Conference on Computational Methods for Coupled Problems in Science and Engineering, Coupled Problems (2013)
4. Birkl, C., Howey, D.A.: Model identification and parameter estimation for *LiFePO*4 batteries. In: Hybrid and Electric Vehicles Conference 2013 (HEVC 2013), p. 2.1–2.1. Institution of Engineering and Technology, Institution of Engineering and Technology, London (2013). doi:10.1049/cp.2013.1889

5. Blochwitz, T., Otter, M., Arnold, M., Bausch, C., Clauß, C., Elmqvist, H., Junghanns, A., Mauss, J., Monteiro, M., Neidhold, T., Neumerkel, D., Olsson, H., Peetz, J.V., Wolf, S.: The functional mockup interface for tool independent exchange of simulation models. In: 8th International Modelica Conference 2011, pp. 173–184 (2011). doi:10.3384/ecp12076173
6. Blochwitz, T., Otter, M., Akesson, J.: Functional mockup interface 2.0: the standard for tool independent exchange of simulation models. In: NAFEMS World Congress (2013)
7. Boland, J., Thibeault, C., Zilic, Z.: Using MATLAB and simulink in a SystemC verification environment. In Proceedings of Design and Verification Conference (2005)
8. Chen, M., Rincon-Mora, G.: Accurate electrical battery model capable of predicting runtime and I-V performance. IEEE Trans. Energy Convers. **21**(2), 504–511 (2006). doi:10.1109/TEC.2006.874229
9. Chen, W., Huhn, M., Fritzson, P.: A generic FMU interface for Modelica. In: 4th International Workshop on Equation-Based Object-Oriented Modeling Languages and Tools, pp. 19–24 (2011)
10. Corbier, F., Loembe, S., Clark, B.: FMI technology for validation of embedded electronic systems. In: Embedded Real Time Software and Systems (2014)
11. Doulos: SystemC Golden Reference Guide. Doulos, Ringwood (2002)
12. Doyle, M., Fuller, T.F., Newman, J.: Modeling of galvanostatic charge and discharge of the lithium/polymer/insertion cell. J. Electrochem. Soc. **140**(6), 1526–1533 (1993). doi:10.1149/1.2221597
13. Elsheikh, A., Awais, M.U., Widl, E., Palensky, P.: Modelica-enabled rapid prototyping of cyber-physical energy systems via the functional mockup interface. In: 2013 Workshop on Modeling and Simulation of Cyber-Physical Energy Systems, MSCPES 2013, pp. 1–6 (2013). doi:10.1109/MSCPES.2013.6623315
14. Functional Mock-up Interface for Co-Simulation, Version 1.0 (2010)
15. Functional Mock-up Interface for Model Exchange and Co-Simulation, Version 2.0 (2014)
16. Grotker, T.: System Design with SystemC. Kluwer Academic, Norwell (2002)
17. He, H., Xiong, R., Guo, H., Li, S.: Comparison study on the battery models used for the energy management of batteries in electric vehicles. Energy Convers. Manag. **64**, 113–121 (2012). {IREC} 2011, The International Renewable Energy Congress
18. ICOS Independent Co-Simulation - User Manual Version 3 (2013)
19. IEEE Standard 1666: SystemC Language Reference Manual (2011)
20. Krammer, M., Karner, M., Fuchs, A.: Semi-formal modeling of simulation-based V&V methods to enhance safety. In: Proceedings of the Embedded World 2014 Exhibition and Conference. WEKA Fachmedien GmbH, Nuremberg (2014)
21. Krammer, M., Karner, M., Fuchs, A.: System design for enhanced forward-engineering possibilities of safety critical embedded systems. In: 17th International Symposium on Design and Diagnostics of Electronic Circuits Systems, pp. 234–237 (2014). doi:10.1109/DDECS.2014.6868797
22. Krammer, M., Fritz, J., Karner, M.: Model-based configuration of automotive co-simulation scenarios. In: Proceedings of the 47th Annual Simulation Symposium. The Society for Modelling and Simulation International, San Diego (2015)
23. Lam, L.: A practical circuit-based model for state of health estimation of li-ion battery cells in electric vehicles. Ph.D. thesis, TU Delft, Delft University of Technology (2011)
24. Lam, L., Bauer, P., Kelder, E.: A practical circuit-based model for Li-ion battery cells in electric vehicle applications. In: 2011 IEEE 33rd International Telecommunications Energy Conference (INTELEC), pp. 1–9 (2011). doi:10.1109/INTLEC.2011.6099803
25. Neema, H., Bapty, T., Batteh, J.: Model-based integration platform for FMI co-simulation and heterogeneous simulations of cyber-physical systems. In: Proceedings of the 10th International Modelica Conference, pp. 235–245 (2014)
26. Newman, J., Thomas-Alyea, K.E.: Electrochemical Systems, 3rd edn. Wiley, Hoboken (2004)
27. Noll, C., Blochwitz, T.: Implementation of modelisar functional mock-up interfaces in SimulationX. In: 8th International Modelica Conference (2011)

28. Piller, S., Perrin, M., Jossen, A.: Methods for state-of-charge determination and their applications. J. Power Sources **96**(1), 113–120 (2001). In: Proceedings of the 22nd International Power Sources Symposium
29. Plett, G.L.: Extended Kalman filtering for battery management systems of LiPB-based HEV battery packs: part 2. Modeling and identification. J. Power Sources **134**(2), 262–276 (2004)
30. Pohlmann, U., Schäfer, W., Reddehase, H., Röckemann, J., Wagner, R.: Generating functional mockup units from software specifications. In: Proceedings of the 9th International MODELICA Conference, 3–5 September 2012, Munich, pp. 765–774 (2012). doi:10.3384/ecp12076765
31. Puntigam, W.: Coupled simulation: key for a successful energy management. In: Virtual Vehicle 11th Automotive Technology Conference (2007)
32. Pussig, B., Denil, J., De Meulenaere, P., Vangheluwe, H.: Generation of functional mock-up units for co-simulation from simulink using explicit computational semantics. In: Proceedings of the Symposium on Theory of Modeling & Simulation - DEVS Integrative, DEVS '14, pp. 38:1–38:6. Society for Computer Simulation International, San Diego (2014)
33. Qtronic: FMU SDK: free development kit (2014)
34. Swan, S.: An introduction to system level modeling in SystemC 2.0. Review Literature and Arts of the Americas (May), pp. 0–11 (2001)
35. SystemC Verification Standard Specification (2003)
36. Unterrieder, C., Huemer, M., Marsili, S.: SystemC-AMS-based design of a battery model for single and multi cell applications. In: 2012 8th Conference on Ph.D. Research in Microelectronics and Electronics (PRIME), pp. 1–4 (2012)

Part II
Reconfigurable Systems and FPGAs

Chapter 3
Building a Dynamically Reconfigurable System Through a High-Level Development Flow

David de la Fuente, Jesús Barba, Julián Caba, Pablo Peñil, Juan Carlos López, and Pablo Sánchez

3.1   Introduction

For nearly two decades, advances on reconfigurable devices have made FPGAs more and more attractive for embedded system designers. Besides the improvement shown in design area, performance, efficiency, and dedicated resources (i.e. DSPs or BRAMs), the most important FPGAs vendors have also included a special characteristic in some of their products: the partial reconfiguration facility or dynamic partial reconfiguration (DPR).

The DPR increases FPGA flexibility because its use is not limited to systems which are resolved at compilation time. A user can modify part of the device configuration at run time, allowing small changes on the initial design after its deployment without a complete configuration process. However, partial reconfiguration is far from being mainstream mainly because of the complexity of its development process [1].

One of the advantages of partial reconfiguration is the system cost reduction due to area reuse. One example of this strategy is software defined radio applications where it is not necessary to deploy all possible radio waveforms at the same time. Another advantage is the flexibility provided by the fact that new components can be developed and instantiated after the initial deployment of the design [2].

D. de la Fuente (✉) • J. Barba • J. Caba • J.C. López
Technology and Information Systems Department, School of Computer Science, University of Castilla La Mancha, Ciudad Real, Spain
e-mail: david.fuente@uclm.es

P. Peñil • P. Sánchez
Electronics Technology, Systems and Automation Engineering Department, Microelectronics Engineering Group, University of Cantabria, Santander, Spain

© Springer International Publishing Switzerland 2016
R. Drechsler, R. Wille (eds.), *Languages, Design Methods, and Tools for Electronic System Design*, Lecture Notes in Electrical Engineering 385, DOI 10.1007/978-3-319-31723-6_3

Therefore, this feature is also of great help when dealing with system maintenance and management of the cycle of life of a product.

However, there are some disadvantages when considering partial dynamic recon-figuration. First, only a few device families implementing run-time reconfiguration are available and, once a manufacturer is chosen, the designer must remain tied to the tools and the workflows imposed by the vendor. Also, partial reconfiguration forces developers to deal with traditional co-design techniques and high-level design problems [3]. In addition, vendor design flows, such as ISE or Vivado design tools from Xilinx, work at a relatively low abstraction level [4].

The work presented in the following sections follows the goal of reducing the complexity of building partially reconfigurable systems by means of a workflow aimed to ease the design of this type of systems. During the design phase, multiple alternatives can be considered and evaluated since it is possible to specify the number and type of functional components deployed in the FPGA fabric. The objec-tive is to analyze the possible implications of each variation in the entire system behaviour. Therefore, different properties such as power consumption, performance or critical path should be considered in the analysis at early stages so that non-promising solutions are avoided as soon as possible, optimizing the efforts. To this end, new design strategies, performing early design space exploration processes have been developed. By applying this approach to partial reconfigurable systems, the search for the best FPGA element configuration is accomplished according to the specific characteristics and requierments that determine the correctness of the system (functional and not functional).

The development of electronic system-level (ESL) design methodologies [5] provides a strategy for designing complex systems, in which the initial key activity is specification. SystemC [6] is the most widely adopted language by the ESL community when it comes to the writing of system executable models. Model-Driven Architecture (MDA) [7] is a design approach that enables expressing systems by means of models at different abstraction levels. As to MDA, the most widely accepted and used language the Unified Modelling Language (UML). Nevertheless, UML lacks of specific semantics required to support all the three embedded system specification, modelling and design steps. In order to overcome this limitations, several application-specific UML profiles have been proposed [8]. In this context, the standard MARTE profile [9] has been developed so that the modelling and analysis of real-time embedded systems is now possible. MARTE provides the concepts needed for capturing real-time features which are the corpus of the semantics in this kind of systems at different abstraction levels. By using this UML profile, designers are able to specify the system functionality composed of interconnected application components and the HW/SW platform where the application is executed. In addition to that, The UML Testing Profile [10] enables the definition of models that capture scenarios for system testing.

Following this combined approach, this paper presents an infrastructure which generates, in an automatic way and using UML/MARTE model as the sole input, the code for both the simulation and implementation of dynamic reconfigurable systems. This infrastructure allows obtaining executable SytemC specifications for simulation and validation purposes according to the system requirements. Finally, our UML/MARTE methodology captures enough information about the target platform so that automatic VHDL code generation of the part to be allocated in the FPGA fabric is also possible.

This chapter is organized as follows; in Sect. 3.2, a study of the state-of-the-art is presented. In Sect. 3.3, the complete design framework is described. Section 3.4 includes a case study that will be used to explained all the aspects of the UML/MARTE methodology. The experimental results are given in Sect. 3.5 and finally, some conclusions are presented in Sect. 3.6.

3.2   Related Works

Generally speaking, the development flow provided to designers in order to build partial reconfigurable systems is complex and difficult. The need, most of the times, of knowing low-level implementation details, may be one the reasons which has prevented this technology from a higher degree of adoption in the industry. The management of the partial reconfiguration process, typically driven through a software routine and a hardware component. Although the flexibility provided by the software, this solution does not give good results, the reconfiguration time is too high [11, 12]. Consequently, these solutions require the use of overlapping techniques between tasks planning and partial reconfiguration, or the prefetching of reconfigurable modules, thus increasing the complexity of tasks management [13, 14].

In summary, the DPR process increases the powerful of the whole system by reusing elements that have been already implemented without redesigning the solution from the scratch, and after the system has been deployed. This characteristic is very convenient for all those applications that require real-time adaptability [15]. The DPR is a useful characteristic under those contexts in which the system cannot be stopped, but it needs to be adapted. Thus, the addition of new hardware components into the system requires that this one is instantiated in a partial reconfigurable region of the FPGA. Due to the fact that a dynamically reconfigurable FPGA might keep running, even when a component failure has been detected, it strengthens the robustness and the reliability of pre-emptive systems. Therefore, a system breakdown can be avoided by replacing the damaged component by a new one at run-time, improving in this way the system fault-tolerance [16].

The characterization of a system specification using a model helps the developer to obtain a global vision of the system to be built, since the model is developed using an approach based on the objectives of the system.

Targeting SystemC [17], enables building executable and platform agnostic validation environments. SystemC is a language which has been widely used for system-level and reusable test bench development and, moreover, the development of advance features intended to support verification and debugging.

Modelling a system brings about many benefits, among which the following can be highlighted: it helps to capture and organize the knowledge of the system, it allows the early exploration of alternatives, it facilitates the decomposition and modularization of the system, it reduces the total number of errors, it facilitates the reuse and maintenance of the system, increasing the productivity of the development team and, finally, it simplifies the documentation process, etc.

Several proposals for bridging the gap between UML specifications and SystemC executable models have been published. The first area where UML and SystemC teamed-up was the use of UML stereotypes for SystemC constructors. This combination focused on a system-on-Chip (SoC) design methodology, such as [19] or [20]. With the work of Riccobene et al. [21] a SoC design flow was proposed based on a SystemC profile aimed at the production of executable models from UML specification.

Most of the efforts spent on the integration of UML into the design process of embedded systems have targeted the synthesis of the hardware and software parts, indistinctly. Several research works on application code synthesis from UML models are characterized by the creation of state machine models or variations of them [22]. In [23], a formal design for reconfigurable, modular digital controller logic synthesis is presented. By means of UML state machines, concurrent super-states are modeled enabling the direct, automatic mapping on structured array of cells in FPGAs.

Nevertheless, the major limitation of these approaches comes from the fact that UML, as a completely generic language, usually lacks of the semantics required to adequately model all the characteristics of embedded systems. In order to confront the challenge and cover the whole design flow of real-time embedded systems, the MARTE profile was created. Taking MARTE-based models as input, several synthesis approaches have also been proposed [9].

MoPCoM [24] is a methodology for the design of real-time embedded systems which supports UML and the MARTE profile for system modelling. Specifically, MoPCoM uses the NFP MARTE profile for the description of real-time properties, the HRM MARTE profile for platform description and the Alloc MARTE profile for architectural mapping.

Gaspard2 [25] is a design environment for data-intensive applications which enables MARTE description of both, the application and the hardware platform, including MPSoC and regular structures. In [26] a generic control semantics for the specification of system adaptability and dynamic reconfigurability in SoCs is presented. The dynamic reconfigurability is implemented by generating the code for a reconfigurable region according to a high level application model, translating it into a hardware component and generating the source code for a reconfiguration controller. The controller is in charge of the management of the different implementations related to the hardware resource. In addition to that, [27] enables the design of FPGA-based embedded system, supporting automatic generation of VHDL descriptions from UML/-MARTE models, establishing a mapping rules to translate high-level elements into VHDL constructs, allowing the generation of fully synthesizable descriptions, including the embedded system structure and behaviour.

3.3   Design Framework

The task of building a reconfigurable project is long and complex mainly due to the currently available tools work at a low level of abstraction. The proposed design framework aims to rise the level of abstraction of the tasks to be performed by the developer. As a consequence, the effort needed to turn a static design into a dynamic one is minimized. Our methodology starts from the creation of a high-level UML/MARTE model of the dynamically reconfigurable system, based on a component-based methodology.

The main advantage of this approach is the use of a standard way of modeling reconfigurable systems, avoiding the development of new profiles; the developer is already overwhelmed by the variety of languages or syntax at his disposal.

After the modeling step, the flow provides the developers with two alternatives: the first one allows to simulate the design by means of a SystemC executable specification (thus, the temporal behavior of the system can be estimated at an early stage of development lifecycle) and; the second one links with the available vendor tools in order to automatically synthesize the bitstreams to configure the reconfigurable fabric. Figure 3.1 illustrates the main steps of the proposed design flow.

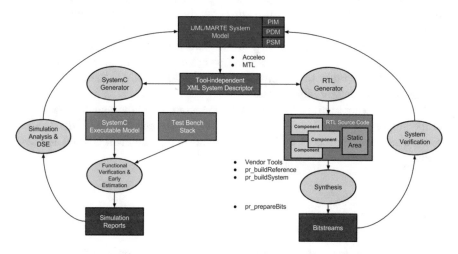

Fig. 3.1 Design flow for dynamically reconfigurable systems design

In order to create the UML/MARTE model, the designer uses the Papyrus [28] tool which is a Eclipse plugin to graphically represent the system. Papyrus is also capable of generating a XMI representation of such model which are the inputs of a XML code generator. This code generator uses Acceleo [29] (a code generation framework fully integrated in Eclipse called) and uses a set of generation templates written in the standard MTL language [30]. The XMI→XML transformation has been done in such a way that other external tools can bind to the development flow in this step.

Once the XML files have been generated, the simulation branch builds a SystemC verification platform which allows the developers to run the SystemC executable specifications, generated from the model information. Depending on the values of each parameter defined in the UML/MARTE model, the number of executable specifications can be variable but all of them will be created automatically. In order to provide the ability of estimating the best temporal setting for a system, when the execution of all specifications are finished, a report is generated showing the configurations ordered according to different criteria. One if the estimated execution time of the application under emulation. Other metrics presented to the designer are the throughput, total traffic of data or bandwidth required by the solution, and the waiting times due to bus congestion or component overloading.

When a system configuration is selected, the synthesis flow can be launched, in this case, the work proposes a solution close to Xilinx development flow. This flow starts with the automatic generation of all the static components to be deployed in the FPGA (i.e. the function that is not going to change). The components which are assigned to dynamic areas—and, therefore, are suitable to be replaced in run-time—must go under a pre-process step so that they are adapted to the system requirements (e.g. type of bus, number of reconfigurable areas, etc.). In the end, three tools (pr_buildSystem, pr_buildReference and pr_prepareBits) are used as front-end to generate the partial bitstream correctly. Each FPGA technology exposes their particularities as to the format and process to (re)program the configuration memories. Therefore, this last step is mandatory to adapt the bitstream to the platform requirements. For example, in our prototype, the bitstream headers had to be modified in order to be compliant with the Xilinx ICAP, the component in charge of the reconfiguration process.

In the following section, the reader will be accompanied through the operational of the proposed design flow. For the sake of simplicity and ease the understanding of the concepts and process, an image processing application will be used as a case study.

3.4 Case Study

In this section we apply the design flow sketched above to generate the implementation code for an image processing application. The main goal of the application is to perform edge detection on an image sequence by applying the SOBEL filter.

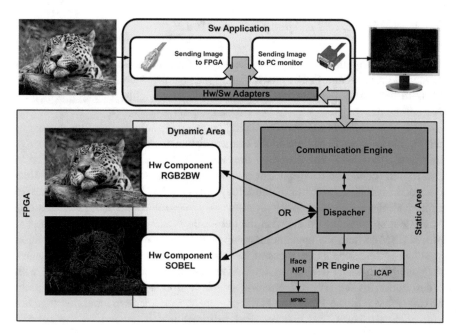

Fig. 3.2 Case study application

The filter requires grayscale pictures in its input, so a previous "RGB to Black and White" transformation is needed. Hence, in the system there will be two components that will be exchanged over the same dynamic area at runtime (i.e. SOBEL and RGB2BW). Finally, the image obtained as the result of the processing chain will be displayed in a computer monitor. All computation tasks will be carried out in an FPGA, in particular a Virtex-5 XC5VFX110T model from Xilinx. In order to receive the images to be processed, the FGPA is connected via Ethernet link to a PC which will send the pictures at a constant rate. This application works with images whose size is 640×480 pixels.

The application (illustrated in Fig. 3.2) will be made up of four components: (a) the first takes a sequence of images from the camera and sends them to the FPGA via Ethernet; (b) the second one converts the image to grayscale; (c) the SOBEL filter will be applied by the third component and finally and; the last component (d) transfers the image to the video memory so it is displayed in a monitor.

Since it is considered a scenario where the available resources are not enough to place the RGB2BW and SOBEL filters, it is necessary to use a dynamic area where one of them will be deployed. Thus the application is able to deploy the functionality that needs at a specific time.

3.4.1 UML/MARTE Specification Model

In this proposal, the goal is to make it possible the automatic generation of executable models and FPGA programming files from component-oriented models [31]. To this end the application of the Model-Driven Architecture [32] principles and techniques to heterogeneous embedded systems is proposed.

However, UML, as a generic modeling language, usually lacks of the semantics required to adequately capture all the characteristics of embedded systems. To solve this issue, the OMG has proposed the MARTE profile [9] as an alternative to confront the challenge that represent covering the complete design flow of real-time embedded systems. Thus, our methodology relies on the UML/MARTE standard to capture all the system details.

System specification starts with the definition of the *the Platform-Independent Model* (PIM) where the application components are modeled, and the application structure is described as instances of these application components who exchanges data using a message delivery system. After that, the model is complemented with a description of the HW/SW platform which conforms the *Platform-Description Model* (PDM). In the PDM, the HW component resources are identified together with their attributes (e.g. the required features) which depend on the specific application under consideration.

Finally, the PIM and PDM models converge into the *Platform-Specific model* (PSM) where a mapping between functional components and available resources in the HW/SW platform is established.

The graphical orientation of UML helps the designers to handle large systems in an easy way. However, the UML/MARTE model must contain all the relevant, essential information, of the system in order to allow the execution of the simulation and synthesis process. The combination of the visual representations with large amounts of data if a powerful tool which could be got using a UML/MARTE methodology, the core of our proposal. In order to handle the design process in the most effective way, the information contained in a UML/MARTE model is separated in specific concerns or domains, depending on their application area. Each concern is captured in a model view, which is represented using the most suitable UML diagrams for a specific concern.

3.4.2 Application Components

Application components identify pieces of functionality that represent a certain behavior with a relevant role in the hierarchy of the system. The entire system is composed of application components which are, in turn, interconnected. This interconnection is established by services, grouped in interfaces. In this methodology, the application components are modeled by the MARTE stereotype <<RtUnit>>. The functionality of each application component is enclosed in two elements:

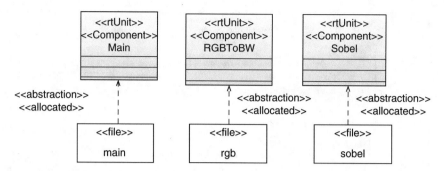

Fig. 3.3 Application components

interfaces and files. The interfaces group the services provided/required needed for establishing the data transmission between the application components. The interfaces are modeled by means of UML interfaces specified by the MARTE <<ClientServerSpecification>> stereotype.

The case study defines three components but only two of them will be implemented in SW (Fig. 3.3). Each application component has an associated set of code files that define the component functionality. These files are modeled as UML artifacts using the UML standard stereotype <<File>>.

The communication is established using two elements: ports and channels. The ports enclose the interfaces that are provided or required by the component. The ports are specified by the MARTE stereotype <<ClientServerPort>>. Each port can support only one interface, and it can be specified with the values provided or required from the attribute kind of the ClientServerPort stereotype, depending on how this interface is used in the component. The system is composed of interconnected instances of application components types (Fig. 3.4). The application instances are connected by links between clients and servers who owns ports implementing a compatible interface. These instances are connected through UML connectors, linking ports (Fig. 3.4) with compatible interfaces.

After that, functional components are then mapped to memory spaces (Fig. 3.5), which will be assigned to HW resources in the target platform model. Thus, components associated with a single memory space are always handled together. Specifically, all the application instances that are realized in HW should be grouped in the same memory space (Fig. 3.5, "rgb2bw" and "sobel"). This association with memory spaces enables the mapping to non symmetric systems, since the model ensures that there are no memory access problems.

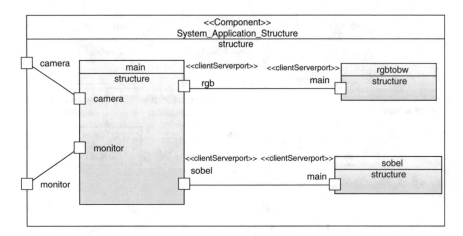

Fig. 3.4 Application system structure

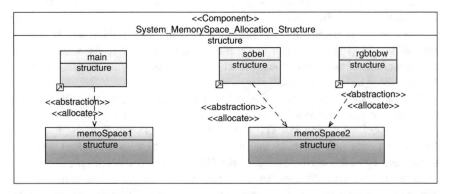

Fig. 3.5 Memory spaces structure and application mapping

3.4.3 HW/SW Platform

The platform where the application is executed includes the HW resources and the SW infrastructure. The hardware platform is captured and supported through MARTE stereotypes of the Hardware Resource Modeling (HRM) subprofile (Fig. 3.6). Processors are modeled with the stereotype <<HwProcessor>>. The bus communication element is also supported through the <<HwBus>> and also different types of memories such as instruction or data caches (stereotype <<HwCache>>) or program memory (stereotype <<HwRAM>>). Each stereotype enables the possibility to set the values of different attributes of the platform.

The other processing element is the FPGA. Considered as a single resource, the FPGA is modeled by the MARTE stereotype <<HwPLD>>. In order to enable

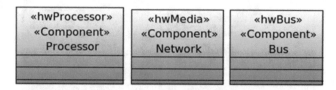

Fig. 3.6 Hw component resources

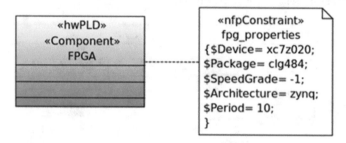

Fig. 3.7 FPGA component

the synthesis process, this component defines compulsory properties for its characterization. These properties are $device, $package, $speedGrade, $architecture and $period (Fig. 3.7).

Additional features have to be specified in the FPGA. Specifically, these features are related to the structure of the FPGA to define the different areas available for configuration. In a composite structure diagram associated to the HwPLD component, the different sections of the FPGA are modeled as parts typed as <<HwComputingResource>> MARTE stereotype. Then, these parts are specified by the MARTE stereotype <<HwComponent>>. In the attribute position, the localization of each section can be specified. Another set of different components considered in the methodology are the resources for communicating the processing elements. Networks and network interfaces can be included in the HW model, with the <<HwMedia>> and <<HwEndPoint>> MARTE stereotypes, respectively (Fig. 3.6). According to the additional properties of the HwEndPoint, different types of communications are captured. When the network interfaces includes the IPAddress property the communication is an IP (Internet Protocol) connection, involving a different synthesis process (Fig. 3.8). When the network interfaces includes the serialPort attribute (Fig. 3.8) the communication is by serial port.

Regarding the SW infrastructure, the operating system is modeled by an UML component tagged with the stereotype <<OS>>[33]. The attributes associated with this stereotype are the type of OS (Ubuntu, Fedora, Angstrom, etc.) and the policy which defines the process scheduling algorithm (Fixed Priority Preemptive, Round Robin, FIFO, EDF, etc.). Finally, when HW and SW components are defined, the

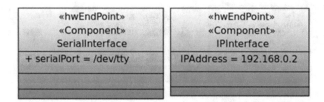

Fig. 3.8 Communication interfaces components

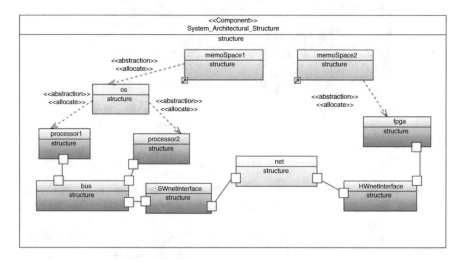

Fig. 3.9 HW/SW platform and memory spaces allocation

target platform is created by using parts typed by the previous HW/SW components (Fig. 3.9). In addition to that, the allocation of the memory spaces previously defined is captured (Fig. 3.9).

3.4.4 Environment

In order to support code generation for evaluation, a testbed has to be integrated. To ease the validation of the design, a set of UML components is specified by means of stereotypes included in the standard profile UML UTP. The components which represent environment elements are specified by the UTP stereotype <<TestComponent>>.

Another component is tagged with the UTP stereotype <<TestContext>>. This component defines the structure of the environment and the interconnection of environmental components with the system under test. The environment structure is modeled in a UML composite structure diagram associated with this *TestContext* component (Fig. 3.10). The composite structure diagram contains instances

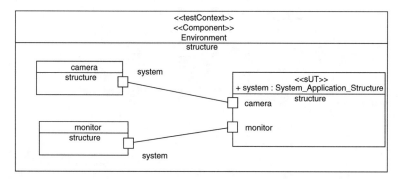

Fig. 3.10 Environment structure

of *TestComponents* previously defined, connected to a UML property typed by
the component where the application is captured (see components defined in
Figs. 3.6, 3.7, and 3.8). This property is specified by the UTP stereotype SUT
(System Under Test) as can be seen in Fig. 3.10.

3.4.5 Simulation Branch

The main goal of this stage is to simulate the final behavior of the system as
soon as possible in the design process. As a result, the designer is provided with
accurate measures about the performance of the system. It is the responsibility of the
system architect to interpret and propose the necessary changes in order to meet the
requirements. As a distinctive feature of this work, the reconfigurable process is also
considered for emulation which gives an enhanced picture of the actual behavior of
the final system.

All this can be achieved keeping the degree of manual intervention in a minimum
level. Therefore, a reconfigurable system architect is only requested to write the
behavior and specify some temporal parameters of the hardware components which
will be deployed in the reconfigurable fabric.

A Reconfigurable Unit (RU) entity is defined as a container for the core function
of an application component. Both elements (RU+core function) results in what is
called a *SystemC application component.* In order to emulate the reconfiguration
process, each SystemC component is compiled as a dynamic library. This way,
the programming of the FPGA, which results in a change of the behavior of a
reconfigurable area, is simulated by means of loading/unloading the appropriate
software library.

The second level of integration has to do with the management of the communication with other components. A Communication Adapter (CA) is responsible
for the adaptation of the low level FIFO interface of the SystemC application
component to TLM semantics. The new logical layer adds additional information

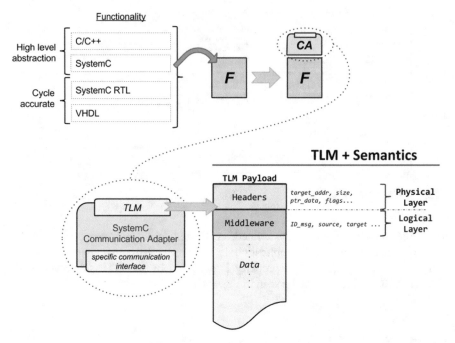

Fig. 3.11 Communication Adapter (CA) scheme

to the messages to be exchanged; identification key of the message, the logical target address or the logical source address, for example. Figure 3.11 graphically represents the composition of the different abstraction layers from the core function to a fully functional system component.

As a result, the simulation infrastructure allows the connection of not only a set of components described in the UML/MARTE, but also other type of systems, as long as the specific CA has been appended. In this way, the capability to communicate with external elements (not necessarily belonging to the simulation environment itself) is achieved, on top of the higher flexibility in the communication topology.

Once the components have been designed, it is necessary the figure of a Reconfiguration Manager (rc_M) to emulate all aspects as to the reconfiguration process such as the interchange of functionality (loading or unloading the dynamic libraries) and management of the execution state of the components (start and stop the process). In addition, the rc_M provides a communication mechanism which allows carrying out the reconfiguration of each component (Fig. 3.12).

After modeling the reconfigurable system using SystemC, a behavior analysis is needed. Taking into account the temporal parameters of the components, reconfiguration latency or dynamic area distribution, the output of this phase provides an advanced knowledge on system behavior. In doing so, we can ensure that the system meets the user requirements. To get it, a testbed module (TBM) has been included in the SystemC infrastructure, previously described (Fig. 3.12).

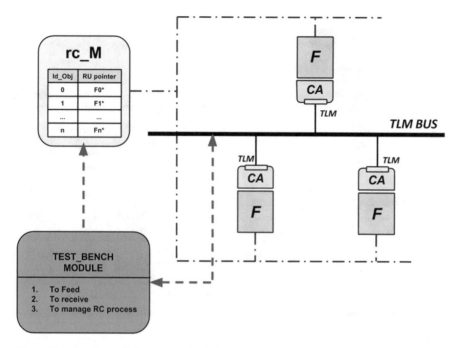

Fig. 3.12 Testbed module for system simulation

TBM offers the ability to inject/receive data to/from the system and manage the reconfiguration process. The actions allow to emulate all possible tasks that may occur in the system, to evaluate its delays and to simulate different scenarios in order to compare different configurations of the system (Design Space Exploration). If the entire system behaves as expected, the following step is to implement the system, starting with the construction of the static area.

3.4.6 RTL Synthesis Branch

Dynamic reconfiguration management is typically solved using a software approach, where an embedded processor is in charge of transferring the reconfigurable bitstream to the configuration memory of the device in order to change the content of the dynamic area. The approach presented in this work, however, relies on a specialized hardware component: the Reconfiguration Engine (referred as prEngine). This has two main advantages: first, it obtains an improvement of two orders of magnitude in speed compared with the software approach [34] and; second, no processor is required for reconfiguration management, freeing CPU clock cycles for other more critical tasks.

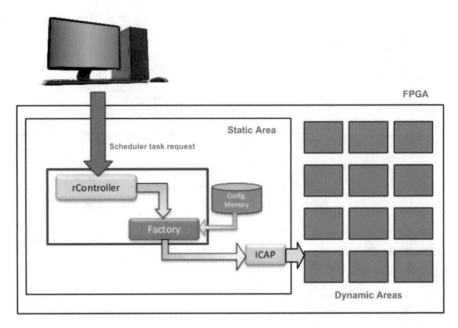

Fig. 3.13 Block diagram of prEngine component

The infrastructure allocated in the static area of the FPGA is composed of the prEngine that offers a set of dynamic reconfiguration related services through a simplified interface. From the architectural point of view, the prEngine has three elements: the Factory, the Reconfiguration Controller and some kind of storage (i.e. external memory), although it is not really part of the engine, and can be shared with the rest of the resources in the system (Fig. 3.13). The Factory component deals with the issues related to the transference and manipulation of partial bitstreams and the Reconfiguration Controller (rController) is responsible for high-level management of the reconfiguration process. It receives the reconfiguration commands and issues the corresponding operation requests to the reconfiguration areas or to the Factory.

The main purpose of the prEngine is to provide a common abstraction layer compatible with all the technologies, by means of encapsulating the mechanism as a transparent service to the upper abstraction layers. Therefore, the prEngine is automatically generated when a dynamic areas is defined in the UML/MARTE specification model. The prEngine structure does not change in any device, however there are some parts which depend on the FPGA technology. In this case, a specific tool from Xilinx called Coregen is used as the backend. A .cgp file is generated from the specification model which is processed by Coregen to build hardware components for a specific device or family. In our proposal, the partial bitstreams are stored in DDR memory so the read and write operations are supervised by the Multi Port Memory Controller (MPMC).

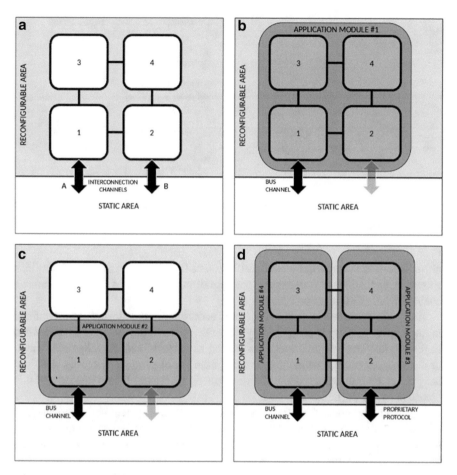

Fig. 3.14 Hierarchical dynamic areas. (**a**) Reconfigurable area structure. (**b**) Single component in four areas. (**c**) Single component in two areas. (**d**) Two components in four areas

Once the static part of the design has been identified, the layout of the dynamic reconfigurable areas must be defined. For each reconfigurable area, at least a geometric location, its resources and a name are specified in the UML/MARTE model. The geometric location delimits a rectangular shape using the bottom-left and top-right slice identifiers.

In the solution proposed, one reconfigurable area is, actually, composed by a grid of smaller sub-areas (see Fig. 3.14a). These sub-areas are interconnected by generic data and control signals which enable the communication between them. Also, a reconfigurable area may have one or more interconnection channels from/to the static area of the FPGA. The designer must enable or disable these interconnections channels and define the type of use they are going to support which can be a bus

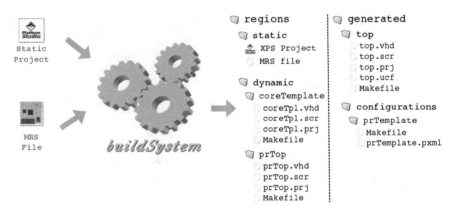

Fig. 3.15 pr_buildSystem tool

or point-to-point protocol, for example. Channel B is disabled in the configurations shown in Fig. 3.14b, c whereas the configuration in Fig. 3.14d has both channels enabled with a different configuration.

A top file described in vhdl is generated according to the dynamic areas defined in the model. This top is composed by the static area and the dynamic areas defined in the model. The dynamic areas are treated as a black box, thus the deployment of any component is possible due to the system is not completely closed. To close the project a configuration definition (described in a *.pxml file) is needed and it is provided by the pr_buildReference.

The next step consists in the construction of the file system and template files of the reconfigurable project. This is performed by the pr_buildSystem tool, that requires as input both the *top.ucf* file and the static part of the project that can be obtained from the UML/MARTE model. The tool builds the wrappers and necessary templates to develop dynamic components and different combinations of reconfigurable modules.

Figure 3.15 shows an example of the directory structure and files generated by this tool. The folder regions contains the differents areas defined in the model, firstly, the static folder matches with the static region and a location file of dynamic areas defined in the model. The dynamic folder is divided into two subdirectories, the first one is a template which is used by the user to define the functionality of each component, and the prTop folder defines a dynamic top whose aim is to build a hierarchy of dynamic regions. The folder generated contains all files generated to run the synthesis process: the top file which instantiates the static and dynamic regions and a configuration template folder to define the configurations, thus the configuration matches a dynamic area to its functionality.

Therefore, after the whole project infrastructure is created, the developer must assign the functionality (already tested) to their right directories when the source code was not included in the model, using the appropriate template. Besides, other configurations can be defined by the user whenever the configuration will be compatible, such as the dynamic area is enough to keep the functionality or the resources used is contained into the area. Moreover, if a new component, not previously designed, is built, it can make use of the generated template.

The next step is to build an initial configuration (reference configuration) by means of pr_buildReference tool. This configuration is used as both the initial design to be loaded in the FPGA, and as a reference for the extraction of the partial reconfigurable areas of other configurations, reducing the synthesis time.

The following step is the synthesis of the reconfigurable Project to get the different bitstreams (one per possible combination of dynamic components). The pr_buildSystem and pr_buildReference tools generated several scripts previously, which help designers to automate this process.

Once all the different bitstreams have been obtained, the last step consists in modifying them so that the factory component can operate with them. Recall that the Factory is the component responsible for the partial reconfiguration of FPGA. The *prepareBits tool* removes the header of the original bitstream and discards the data located after desynchronization command. Thus, the valid data is retrieved to build the new bitstream, which is divided into three parts: a reduced header, which includes configuration and synchronization commands, the valid bitstream data, and the desynchonization command. All these commands are aligned to 32 bits words, so that the Factory component gets a high performance using the maximum bandwidth allowed by the ICAP. The tool is also able to meld several partial bitstreams for deploying several components in the same process. In this case the bitstream data matches with the valid data of each original bitstream.

Finally, the verification of the system is needed to guarantee the correctness of the project, including all process: reconfiguration, application deployed into a dynamic area, etc., so the user has to define integration tests to verify the entire system. Due to the fact that some synthesis process are beyond our control, the system can be synthesized in a not desired way, such as timing violations. In that case it might be necessary to make modifications to the reference UML/MARTE model.

3.5   Results

The application described in the case study has been modeled using the UML/-MARTE approach presented followed in previous sections. It includes the parameters and theirs values required for the different SystemC simulation scenarios and the and subsequent synthesis phase.

From the UML/MARTE model, the SystemC simulation process builds the entire simulation infrastructure allowing the temporal evaluation of all possible configuration variations for the reconfigurable system. In this case, the components have associated an execution time estimation required in the simulation phase; the partial reconfiguration process estimation has been acquired through the Factory characterization and depends on the dynamic component which will be deployed too, whereas, the estimates of hardware components are acquired by the High-Level Synthesis tool.

After analyzing the results at this stage, from the simulation log file we can extract the following relevant information:

- The total number of simulated scenarios was 258. This figure comes from the combination of different alternatives related to the type of bus used, burst sizes, size of the memories used, number and structure of the component's internal buffers, etc.
- If only one filter is used, the estimated performance is 0–63 frames per second for the RGB2BW case or 0–59 frames per second for the SOBEL case.
- Using dynamic reconfiguration and both filters in the processing chain, the performance sees a reduction of 35.27 %.
- The bandwidth achieved for reconfiguration purposes (prEngine→ICAP) is 180.3 MB/s.

Once the best temporal configuration is selected from the simulation branch, the synthesis of the application has been carried out, targetting a Virtex5 XC5VFX110T model of Xilinx. The total resources used in the device are shown in Table 3.1.

Table 3.1 Resources in FPGA

Number of BUFGs	4 out of 32	12,5 %
Number of external IOBs	16 out of 640	$\simeq 2,5\%$
Number of LOCed IOBs	4 out of 4	100 %
Number of RAMB18X2s	6 out of 148	$\simeq 4\%$
Number of slices	698 out of 17280	$\simeq 4\%$
Number of slice registers	972 out of 69120	$\simeq 1,5\%$
Number of slice LUTS	1496 out of 69120	$\simeq 2\%$
Number of slice LUT-flip flop pairs	1976 out of 69120	$\simeq 3\%$

As it can be seen in Table 3.1, the overhead imposed by our reconfiguration system is minimal, allowing higher occupancy of the resources of the FPGA by the developer design. Apart from this, in order to get an idea of the benefits of using the proposed design flow, the Table 3.2 shows the lines of code that have been generated automatically instead of handwritten.

Table 3.2 Source code generated

Elements (VHDL)	Lines of code	Elements (SystemC)	Lines of code
Static area	4568	SystemC platform	950
RGB2BW	1483	RGB2BW	69
SOBEL	1771	SOBEL	69
Total	7822	Total	1088

It is worth noticing that nearly 8.000 lines of VHDL source code were generated in total, for developers only have to focus their efforts in the functionality of the system. In this scenario, both HW filters (RGB2BW and SOBEL) were generated from C/C++ using a tool of Xilinx called Vivado HLS. The RGB2BW C++ description only has 15 lines and the SOBEL algorithm 16 lines of source code. With just a few lines, the Vivado HLS tool generates 3254 lines of code, which are integrated with the rest of the system thanks to the envisioned integration infrastructure, methodology and tools.

3.6   Conclusion

This work provides a full Eclipse integrated design flow that allows to create dynamic partial reconfigurable applications from a high abstraction level (UML/-MARTE model), to simulate different scenarios based-on design space exploration and to implement the final application over a FPGA device.

An hierarchical approach to the reconfiguration problem is taken, where top-level reconfigurable areas can be divided into a set of smaller independent regions, that can also be freely composed at run time. The tools rely on the use of a pr_Engine which is provided as an independent IP core. The core provides a common abstraction to the reconfiguration process which is device independent. New FPGA families could be supported with little effort.

Acknowledgements This work has been partly funded by the Spanish Ministry of Economy and Competitiveness under project REBECCA (TEC2014-58036-C4-1-R) and by the Regional Government of Castilla-La Mancha under project SAND (PEII_2014_046_P).

References

1. Bobda, C.: Introduction to Reconfigurable Computing: Architectures, Algorithms, and Applications. Springer Publishing Company, New York, Incorporated (2007). ISBN: 1402060882, ISBN: 9781402060885
2. Dye, D.: Partial Reconfiguration of Xilinx FPGAs using ISE Design Suite (UG702). Xilinx (2011)
3. Wolf, W.: High-Performance Embedded Computing: Architectures, Applications, and Methodologies. Princeton University, Princeton (2006)
4. Xilinx: Partial Reconfiguration User Guide. Xilinx (2011)
5. Steiner, N., Wood, A., Shojaei, H., Couch, J., Athanas, P., French, M.: Torc: towards an open-source tool flow. In: Nineteenth ACM/SIGDA International Symposium on Field-Programmable Gate Arrays (2011)
6. Cervero, T., Lopez, S., Sarmiento, R., Frangieh, T., Athanas, P.: Scalable models for autonomous self-assembled reconfigurable systems. In: International Conference on ReConFigurable Computing and FPGA (2011)
7. Martin, G., Bailey, B., Piziali, A.: ESL Design and Verification: A Prescription for Electronic System Level Methodology (Systems on Silicon) (2007). ISBN-10: 0123735513

8. Vanderperren, Y., Mueller, W., Dehaene, W.: UML for electronic systems design: a comprehensive overview. Des. Autom. Embed. Syst. **12**(4), 261–292 (2008)
9. OMG: UML Profile for MARTE. www.omgmarte.org (2013)
10. OMG: UML Testing Profile (UTP) 1.1. http://utp.omg.org/ (2014)
11. Flynn, A., Gordon-Ross, A., George, A.D.: Bitstream relocation with local clock domains for partially reconfigurable FPGAs. In: Proceedings of the Conference on Design, Automation and Test in Europe (DATE) (2009)
12. Hubner, M., Gohringer, D., Noguera, J., Becker, J.: Fast dynamic and partial reconfiguration data path with low hardware overhead on Xilin FPGAs. IEEE International Symposium on Parallel and Distributed Processing (2010)
13. Noguera, J., Badia, R.M.: Multitasking on reconfigurable architectures: microarchitecture support and dynamic scheduling. ACM Trans. Embed. Comput. Syst. **3**(2), 385–406 (2004)
14. Redaelli, F., Santambrogio, M.D., Sciuto, D.: Task scheduling with configuration prefetching and anti-fragmentation techniques on dynamically reconfigurable systems. In: Proceedings of the Conference on Design, Automation and Test in Europe (DATE) (2008)
15. Viswanathan, V., Atitallah, R.B., Dekeyser, J.L.: Dynamic reconfiguration of modular I/O IP cores for avionic applications. Conference on Reconfigurable Computing and FPGAs (ReConFig) (2012)
16. Dondo, J.D., Rincon, F., Valderrama, C., Villanueva, F.J., Caba, J., Lopez, J.C.: Facilitating preemptive hardware system design using partial reconfiguration techniques. Sci. World J. (2013)
17. IEEE Std. 1666-2011: IEEE Standard for Standard SystemC®Language Reference Manual (2012). Available at http://standards.ieee.org/getieee/1666/download/1666-2011.pdf
18. Vanderperren, Y., Mueller, W., Dehaene, W.: UML for Electronic Systems Design: a comprehensive overview. J. Des. Autom. Embed. Syst. **12**, 261–292 (2008)
19. Bruschi, F., Di Nitto, E., Sciuto, D.: SystemC Code Generation from UML Models. Forum on Specification and Design Languages Best of FDL'02. Kluwer Academic, Boston/Dordrecht/London (2003)
20. Muller, W., et al.: The SATURN approach to sysML-based HW/SW codesign. IEEE Annual Symposium on VLSI, ISVLSI (2010)
21. Bocchio, S., Riccobene, E., Rosti, A., Scandurra, P.: A SoC design flow based on UML 2.0 and SystemC. In: DAC, Workshop UML-Sock'05
22. Harel, D., Kugler, H., Pnueli, A.: Synthesis revisited: generating statechart models from scenario-based requirements. Formal Methods in Software and System Modeling. Springer, Heidelberg (2005)
23. Adamski, M.: Design of reconfigurable logic controllers from hierarchical UML state machines. 2009 4th IEEE Conference on Industrial Electronics and Applications (ICIEA) (2009)
24. Vidal, J., de Lamotte, F., Gogniat, G., Soulard, P., Diguet, J.P.: A code-design approach for embedded system modelling and code generation with UML and MARTE". In: Proceedings of the Conference on Design, Automation and Test in Europe (DATE). Dresden (2009)
25. Piel, E., Atitallah, R., Marquet, P., Meftali, S., Niar, S., Etien, A., Dekeyser, J.-L., Boulet, P.: Gaspard2: from MARTE to SystemC Simulation. In: Proceedings of the DATE'08 workshop on Modeling and Analysis of Real-Time and Embedded Systems with the MARTE UML profile (2008)
26. Quadri, I.R., Huafeng, Y., Gamatie, A., Rutten, E., Meftali, S., Dekeyser, J.-L.: Targeting reconfigurable FPGA based SoCs using the UML MARTE profile: from high abstraction levels to code generation. Int. J. Embed. Syst. **4**(3–4), 204–224 (2010)
27. Leite, M., Vasconcellos, C.D., Wehrmeister, M.A.: Enhancing automatic generation of VHDL descriptions from UML/MARTE models. In: 12th IEEE International Conference on Industrial Informatics (INDIN) (2014)
28. Papyrus webpage: http://www.papyrusuml.org
29. Acceleo webpage: http://www.acceleo.org

30. MOF Model To Text Transformation Language webpage: http://www.omg.org/spec/MOFM2T/1.0
31. Szyperski, C.: Component Software: Beyond Object-Oriented Programming. Addison-Wesley Professional, Boston (2002)
32. Schmidt, D.C.: Model-driven engineering. IEEE Comput. **39**(2), 25–31 (2006)
33. Posadas, H., Peñil, P., Nicolás, A., Villar, E.: Automatic synthesis of embedded SW for evaluating physical implementation alternatives from UML/MARTE models supporting memory space separation. Microelectron. J. **45**(10), 1281–1291 (2014)
34. Dondo, J.D., Barba, J., Rincón, F., Moya, F., López, J.C.: Dynamic objects: supporting fast and easy run-time reconfiguration in FPGAs". J. Syst. Archi. **59**, 1–15 (2013)

Chapter 4
A Special-Purpose Language for Implementing Pipelined FPGA-Based Accelerators

Cristiano B. de Oliveira, Ricardo Menotti, João M.P. Cardoso, and Eduardo Marques

4.1 Introduction

FPGA (Field-Programmable Gate Array) devices allow the development of high-performance digital circuits that can be used for High-Performance Computing (HPC) applications. They provide a high amount of reconfigurable hardware resources, which can be used for approaches that focus on efficient schemes of parallelism in order to accomplish efficient and fast hardware implementations. For such achievement, however, the developer needs a high level of expertise in order to take advantage of the FPGA capabilities, once the efficient use of parallelism is a complex task, specially when an optimized use of the FPGA resources is also required.

A common approach regarding the use of FPGAs is the design of hardware accelerators, which are components intended for speeding up specific computing

C.B. de Oliveira (✉)
Institute of Mathematical and Computer Sciences, Universidade Federal do Ceará (UFC),
São Paulo, Brazil
e-mail: cbacelar@icmc.usp.br

R. Menotti
Departamento de Computação, Universidade Federal de São Carlos (UFSCar), São Carlos, Brazil
e-mail: menotti@dc.ufscar.br

J.M.P. Cardoso
Faculty of Engineering of University of Porto, Universidade do Porto, Faculdade de Engenharia
(FEUP), Porto, Portugal
e-mail: jmpc@acm.org

E. Marques
Institute of Mathematical and Computer Sciences, Universidade de São Paulo (USP),
São Paulo, Brazil
e-mail: edumarques@usp.br

© Springer International Publishing Switzerland 2016
R. Drechsler, R. Wille (eds.), *Languages, Design Methods, and Tools
for Electronic System Design*, Lecture Notes in Electrical Engineering 385,
DOI 10.1007/978-3-319-31723-6_4

tasks. Hardware Description Languages (HDLs), such as VHDL and Verilog, can be used for specifying such components. The use of such languages implies a hardware-oriented level of abstraction, which forces the developer to manage issues like clock cycles, signals and events. In the sense of providing a higher abstraction level for hardware specification, tools for automatic compilation of software programs into FPGA hardware descriptions have been developed. High-Level Synthesis (HLS) tools are meant to significantly decrease the programming efforts required for embedded system developers by allowing them to generate hardware components from high-level languages, such as C and Java.

Despite the easiness of use provided by HLS tools, their presented solutions are many times considered inefficient when compared to the ones achieved by specialized hardware designers, as some implementation aspects, such as parallelism, are not always well handled by automatic compiler techniques. In this regard, Domain-Specific Languages (DSLs) can provide higher abstraction levels than HDLs while giving to the programmer means to express specialized knowledge and to control some issues about the hardware generation process that can lead to better solutions. Figure 4.1 illustrates a general comparison between HDLs, DSL and HLS for hardware accelerators development.

This chapter presents the current version of a DSL named LALP (Language for Aggressive Loop Pipelining) [1], which is a special-purpose DSL for handling parallelism by applying loop pipelining techniques to a C-like source code. The presented LALP features also include a distributed pipelining control, floating-point support and custom memory organizations. The main intention of this chapter is to show possible usages for the language and also differences and advantages of using LALP. With LALP we expect to provide a language that covers portions of the design space for developing FPGA not currently target by other approaches.

This chapter is organized as follows. Section 4.2 describes the LALP language, presenting the overall concept about designing accelerators using LALP. A more detailed description of LALP features and some examples of LALP code are shown in Sect. 4.3. Section 4.4 presents the mains approaches used by the LALP compiler for hardware generation. Experimental results for different LALP architectures are presented in Sect. 4.5, including results for four implementations of a Sobel operator. Section 4.6 describes the most relevant related work. Finally, Sect. 4.7 concludes the paper and describes some of the ongoing work and future work planned.

Fig. 4.1 General comparison between HDLs, DSL and HLS for hardware accelerators development

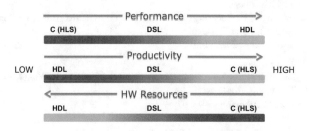

4.2   LALP: Language for Aggressive Loop Pipelining

LALP (Language for Aggressive Loop Pipelining) [1] is a special purpose language for developing hardware accelerators in FPGAs. It is mainly focused on mapping loop structures into reconfigurable hardware descriptions, since critical application sections mainly consist of loops. With this respect, LALP allows developers to implement highly parallel hardware applications by handling loop iterations in pipelining, where different operations of successive iterations of the loop are processed at the same time. LALP facilitates the development of highly efficient hardware accelerators since it supports the control of the execution of each pipelining stage in a higher abstraction level than HDLs.

Figure 4.2 shows the compilation flow used for generating VHDL code from LALP code. LALP relies on a custom hardware components library for VHDL generation. Based on low-level directives that can be expressed by the programmer operations scheduling is performed by the compiler. Such directives are also used for synchronizing signals and to control the dataflow. A modified version of the ASAP/ALAP schedule is used in order to reduce the distances in the back edges and hence the initiation interval between iterations. As output, the compiler generates VHDL code and graph representations of the hardware (e.g., *.dot files). It is also possible to retrieve a structural description of the hardware in a structural LALP language variant named LALP-S [1].

The compiler uses a conservative algorithm to avoid loss of data due to differences on the scheduling of several datapaths. The scheduling relies on the use of counters, which results on an invariable number of cycles per iteration. Despite not supporting the use of irregular loops, this allows the compiler to identify if the operations bound to counters are balanced. The number of clock cycles for each iteration is determined by computing backward edges for loop dependency analysis.

The hardware component library used by the LALP compiler does not explicitly take into account specific FPGA resources. It is specified in generic VHDL code,

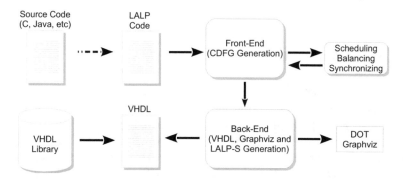

Fig. 4.2 LALP compilation flow

thus, it can be synthesized to different FPGAs (such as the ones from Xilinx® and from Altera®). The LALP compiler relies on the capability of RTL synthesis tools to target efficiently the resources of a specific FPGA.

The main approach used by LALP is to minimize the latency in loop iterations using cascades of registers for balancing and synchronizing the datapath operations. In order to implement this approach, it is necessary to control the number of clock cycles for each operation. LALP uses a VHDL component library that provides counters and shift registers that allow such level of control.

The LALP language intends to allow developers to design accelerators with a high parallelism level. All the statements are by default considered concurrent and the presence of dependencies determines the execution sequence of each statement and the operation scheduling. The dependencies are automatically handled by the LALP compiler which will also schedule the operations according to the delay values defined in LALP code by the use of the qualifier '@'. This qualifier can be used in combination with other directives provided by LALP (e.g., when, @) to setup the execution flow by specifying delay values in LALP code for the components (i.e., counters, shift registers, etc) that will control the pipeline.

By using such directives, programmers can manually schedule the operations of different loop iterations in a pipelined execution since they will define the number of clock cycles for a statement (or operation) to be enabled. Once is possible to specify different values for the clock cycles in each operation, the amount of clock cycles for the whole pipeline changes according to the user-specified delays. This feature allows a high control of the program execution flow.

The manual annotation of scheduling using '@' is optional. In the case where there are no annotations in the LALP code, the compiler tries to schedule the operations and to determine the '@' values needed. For instance, assuming that a counter is used for addressing one array in the beginning of the datapath and other in the end, the set of in-between components will determine the number of clock cycles required for scheduling these accesses. However, if one memory needs to be accessed more than one time per iteration, the developer needs to manually specify the scheduling for such operations.

4.3 LALP Features

A LALP code has a C-like syntax and provides a higher-level of abstractions than using HDLs. However, some hardware aspects are still explicit in order to allow a better design control in comparison to a C-to-gates approach.

A general structure of a LALP source file is shown in Fig. 4.3. Typically, a LALP code has three main sections. The first section is used to define constants and custom data types (lines 1–4), which can be derived from the ones showed in Table 4.1. The second section (lines 6–14) is where the program code is written and, therefore, where the developer specifies the scheduling through the insertion of delays in the pipeline. This is the only section required, since it holds the main logic of the module

```
1  // Constants declarations and types definitions
2  const <identifier> = <value>;
3  typedef <type>(<bits>) <identifier>;
4
5  // Module implementation
6  <identifier>(<ports>)
7  {
8      // Variables
9      {
10             ...
11     }
12
13     ...
14 }
15
16 // Testbench specification
17 assert
18 {
19         ...
20 }
```

Fig. 4.3 General LALP code structure

Table 4.1 Supported types in LALP

Type	Description	Bit width
signed	Signed integer	32 *bits*
unsigned	Unsigned integer	32 *bits*
sfixed	Signed fixed-point	32 *bits* (16 *bits* frac)
ufixed	Unsigned fixed-point	32 *bits* (16 *bits* frac)
float	Floating-point	32 *bits* (Single precision)
bit	Bit/Signal	1 *bit*
signed(N)	Custom *signed*	N bits
unsigned(N)	Custom *unsigned*	N bits
sfixed(M, N)	Custom *sfixed*	$(M + N)$ *bits*, N *bits* frac
ufixed(M, N)	Custom *ufixed*	$(M + N)$ *bits*, N *bits* frac
float(M, N)	Custom *float*	$(M + N)$ *bits*, $(M - 1)$ *bits* for mantissa and N *bits* for exponent

and also the ports for external interfacing. The file can also hold an optional third section, which is meant for asserting the generated hardware according to a test bench specification (lines 17–20).

A complete example of a LALP code for multiplying two arrays is shown in Fig. 4.4. Lines 1–5 declare the constants, custom types and the bit width. In this example the type *int* is defined as signed and as 16 bits (line 5). The module *arr_mult* is declared with one input and one outputs (line 7). The variables are declared in lines 11–14, with x, y and *result* declared as arrays. The addresses of the arrays are defined by the value of the counter i (lines 18, 19, and 20). Figure 4.5 shows the equivalent C code.

```
1  //Constant declaration
2  const N = 3;
3
4  // Type and bitwidth declaration
5  typedef signed(16) int;
6
7  arr_mult(out bit done, in bit init)
8  {
9      // Variables declarations
10     {
11         int x[N]={4,5,6};
12         int y[N]={1,2,3};
13         int result[N];
14         int res,i;
15     }
16
17     counter (i=0; i<N; i++);
18     x.address = i;
19     y.address = i;
20     result.address = i;
21
22     res = x * y;
23     result.data_in = res;
24
25     done = i.done;
26 }
```

Fig. 4.4 Simple LALP code example

```
1  #define N 3
2
3  void arr_mult(int *result, int *x, int *y)
4  {
5      int res,i;
6
7      for (i=0; i < N; i++)
8      {
9          res = x[i] * y[i];
10         result[i] = res;
11     }
12 }
```

Fig. 4.5 C code representing the LALP code in Fig. 4.4

In this example there are no manually inserted delays and the compiler performs the whole scheduling process. Figure 4.6 shows the CDFG (Control Data Flow Graph) for this example, where one can see the delays automatically inserted by the compiler. In this case, the delays are used for control signals, i.e, address and write enable. It is also noticeable that the operations are scheduled in order to balance the datapath. This results in the schedule shown in Fig. 4.7.

Although not required, the programmer has several ways for using delays in this example. For that it would be necessary to change some lines of code. Figure 4.8 shows examples of modified lines of the original code, where delays are used for different purposes, such as delaying the multiplication result (Fig. 4.8a), the memory addressing (Fig. 4.8b) or even to increase the number of cycles required for each

Fig. 4.6 CDFG representing
the LALP code in Fig. 4.4

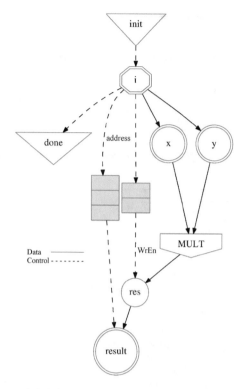

Fig. 4.7 Example of
automatic scheduling,
without use of delays

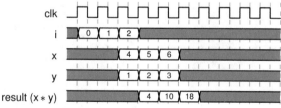

step of the counter (Fig. 4.8c). These cases illustrate some possibilities for using
the qualifier '@'. Each of them produces a different effect in the pipelining. The
corresponding effect for each of these cases is presented in Fig. 4.9. Notice that
some signals are omitted for the purpose of simplification.

A more complex example of LALP code is shown in Fig. 4.10. It is a possible
implementation of a Sobel filter in LALP where the internal loop for computing
the masks is unrolled. In this case, the position of the array *img* is defined by the
variable *addr* (line 26). However, since the array address is tied to the counter, it
is necessary to allow 8 reads from the memory before changing the counter value.
This is done by delaying the counter step during 8 clock cycles (line 16). Thus,
one address is computed for each clock cycle (lines 18–25). For reading each value
from memory in syncing with the addressing we need to setup the assignments of

```
21   ...
22     c = (x *@2 y);
23   ...
```

(a)

```
17   ...
18     x.address = i@2;
19     y.address = i@2;
20     result.address = i@2;
21   ...
```

(b)

```
16   ...
17     counter (i=0; i<N; i++@2);
18   ...
```

(c)

Fig. 4.8 Examples of applying delays in LALP: (**a**) Delay in multiplication; (**b**) Delay in reading the counter value; (**c**) Delay in the increment of the counter value

Fig. 4.9 The impact on execution of applying delays in LALP: (**a**) Fig. 4.8a; (**b**) Fig. 4.8b; (**c**) Fig. 4.8c

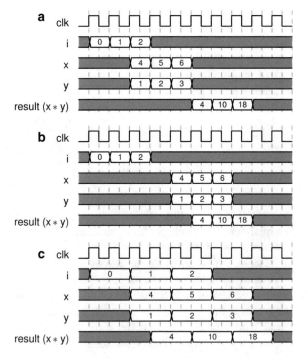

```
 1  const ROWS = 10;
 2  const COLS = 10;
 3  const SIZE = ROWS*COLS;
 4  const N = (COLS*(ROWS-2)) -2;
 5
 6  sobel(in bit init, out bit done, out int result)
 7  {
 8    {
 9      unsigned(16) H, O, V, Hpos, Vpos, res;
10      unsigned(8) i, addr;
11      unsigned(8) i00, i01, i02, i10;
12      unsigned(8) i12, i20, i21, i22;
13      unsigned(8) img[SIZE] = {...}; //omitted
14      unsigned(8) output[SIZE];
15      bit done_ok = 1;
16    }
17    i.clk_en = init;
18    counter (i=0; i < N; i++@8);
19
20    addr = i;
21    addr = (i) + 1 when i.step@1;
22    addr = (i) + 2 when i.step@2;
23    addr = (i) + COLS when i.step@3;
24    addr = (i + COLS) + 2 when i.step@4;
25    addr = (i + COLS) + COLS when i.step@5;
26    addr = ((i + COLS) + COLS) + 1 when i.step@6;
27    addr = ((i + COLS) + COLS) + 2 when i.step@7;
28    img.address = addr;
29
30    i00 = img when i.step@2;
31    i01 = img when i.step@3;
32    i02 = img when i.step@4;
33    i10 = img when i.step@5;
34    i12 = img when i.step@6;
35    i20 = img when i.step@7;
36    i21 = img when i.step@8;
37    i22 = img when i.step@9;
38
39    H = ((-i00) + (-i01 -i01)) + (((-i02) + i20) + ((i21 + i21) + i22));
40    V = ((-i00) + i02) + (((-i10 -i10) + (i12 + i12)) + ((-i20) + i22));
41
42    Hpos = H < 0 ? -H : H;
43    Vpos = V < 0 ? -V : V;
44
45    O = Hpos + Vpos;
46
47    res = O > 255 ? 255: O;
48
49    output.address = i;
50    output.data_in = res;
51
52    result = output;
53    done = done_ok;
54  }
```

Fig. 4.10 Example of LALP code for a Sobel filter

the corresponding registers by using the conditional keyword *when* (lines 28–35). The output value of the filter is computed in lines 37–43. This value will be trunked if greater than 255 (line 45) and this condition is implemented by using a ternary operator, since there are no *if then else* statements in LALP.

```
1  const COLS = 10;
2  const ROWS = 10;
3  const COLS_minus_one = (COLS − 1);
4  const ROWS_minus_one = (ROWS − 1);
5
6  sobel(out bit done, out unsigned(9) result)
7  {
8    {
9      unsigned(16) H, V, O, Hpos, Vpos;
10     unsigned(8) i00, i01, i02, i10;
11     unsigned(8) i12, i20, i21, i22;
12     unsigned(8) img[ROWS][COLS] = {...}; //omitted
13     unsigned(8) output[ROWS][COLS];
14     unsigned(8) x=0, y=0;
15     unsigned(9) res;
16     bit done_ok = 1;
17   }
18
19   counter[ (x = 1; x < ROWS_minus_one; x++)
20            (y = 1; y < COLS_minus_one; y++)];
21
22   i00 = img[x−1][y−1];
23   i01 = img[x−1][y];
24   i02 = img[x−1][y+1];
25   i10 = img[x][y−1];
26   i12 = img[x][y+1];
27   i20 = img[x+1][y−1];
28   i21 = img[x+1][y];
29   i22 = img[x+1][y+1];
30
31   H = ((-i00) + (-i01 -i01)) + (((-i02) + i20) + ((i21 + i21) + i22));
32
33   V = ((-i00) + i02) + (((-i10 -i10) + (i12 + i12)) + ((-i20) + i22));
34
35   Hpos = H < 0 ? −H : H;
36   Vpos = V < 0 ? −V : V;
37
38   O = Hpos + Vpos;
39   res = O > 255?255:O;
40
41   output[x][y] = res;
42
43   result = res;
44   done = done_ok;
45 }
```

Fig. 4.11 Example of Sobel filter with an alternative LALP syntax

As can be noticed in the presented Sobel example, a major difficulty for using LALP is related to the use of '@', specially when handling multiple memory operations. For this reason, LALP also provides a syntax version where the addressing is done by using array index, in a more friendly approach. Figure 4.11 shows a Sobel implementation using this new syntax. Some noticeable differences related to the first version are the use of a nested counter (line 19) and the direct addressing by using indexes (lines 21–28 and 39).

When using this approach some issues related to the memory scheduling and syncing control need to be handled by the compiler. For such purpose, LALP uses custom memory architecture. More details about this are shown in Sect. 4.4.

4.4 Hardware Generation

For hardware generation LALP uses a modified version of the Nenya VHDL Library [2], which contains a large set of VHDL components. The operators for supporting floating-point are implemented in a different custom library.

The main component used by the compiler is the counter that is used in combination with the shift registers for loop control. The used library provides a version of the counter component for single loops. However, nested loops are also supported. In this case, a nested version of the counter is generated during the compilation.

4.4.1 Floating-Point Support

LALP supports basic floating-point (FP) operations: addition, subtraction, multiplication and division. This is done by using VHDL-RTL descriptions for the operators, which were implemented using David Bishop's IEEE Proposed library.[1] The operators consist of 32 bits components, implemented according to IEEE 754 [3].

A second option to support FP is the use of operators specified as LALP modules and using those modules in the code of the application. This approach allows the compiler to generate hardware taking into account specific parameter values that control the generation of the hardware for the operators. Such feature can also be achieved by parameterized VHDL-RTL descriptions or/and circuit generators, but using LALP the dataflow graph representing the operators can be exposed (inlined or not) to the dataflow graph of the design and will make possible to specialize even more the hardware generation as optimizations can be applied to all design levels.

Parameterized LALP code can be used to explore the number of pipelining stages in FP operators in order to increase circuit's clock frequency. Therefore, the modules are implemented with a number of *pipeline* stages according to the operation performed. Thus, it is possible to use LALP code to exploit the number of clock cycles spent in each stage in order to provide the best design for each accelerator.

The two LALP modules available are the ones for addition and multiplication. The stages for each module are defined in Table 4.2. Since these modules are not used by default the programmer needs to set up the appropriated compilation parameter in order to use them.

---

[1]Freely available in http://www.eda.org/fphdl/.

Table 4.2 Pipeline stages used in LALP modules for multiplication/aqd-dition in floating-point (single precision)

	Multiplication		Addition
Stage	Description	Stage	Description
1	Read input values for sign, exponent and mantissa	1	Read input values for sign, exponent and mantissa
2	Compute output sign and exponent	2	Determine the greater exponent
3	Insertion of implicit bit for mantissa	3	Exponents subtraction
4	Mantissa alignment	4	Mantissa alignment
5	Multiplication of mantissas	5	Addition of mantissas
6	Overflow adjustment and number composition	6	Overflow adjustment and number composition

4.4.2 Memory Architecture

The main issue when handling array accesses in LALP is to correctly synchronize multiple reading and writing operations with all the other stages of the pipeline. Each array in LALP is mapped to a single BRAM and, typically, each memory reading or writing demands one clock cycle, but two clock cycles are required for the data to be available after a writing, once the current version of LALP does not support dual-port BRAMs. Besides that, the developer needs to take into account the number of clock cycles for every other operation performed along the datapath. Such concern increases the difficulty on using LALP for implementing very complex accelerators.

The provided solution for this problem is to use the alternative version of LALP, presented in Sect. 4.3. When using this version, the compiler schedules the pipeline considering that the data will be available at the same time for each datapath. Thus, the compiler generates specific architectures to suit this condition. For now, only cases where the input and output are mapped into separated memories are supported.

Different approaches to control the memory accesses can be used according to compilation parameters. Figure 4.12 shows the default architecture generated by the compiler in the cases where more than one memory accesses are needed. In such case, the counter step is delayed according to the number of memory addresses handled by loop iteration. For instance, for reading four addresses the counter value will change after four clock cycles and one memory position will be accessed during each of them. An automatically generated Finite State Machine (FSM) controls the syncing in order to ensure that the data will be processed in each datapath in the correct time.

LALP uses an approach intrinsically based on counters, tapped delay lines, and distributed control. Thus, the use of FSMs is limited to enable pipeline stages

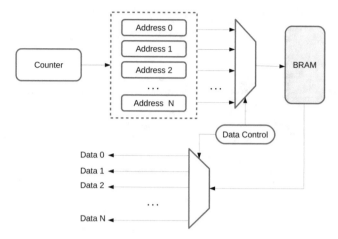

Fig. 4.12 Architecture used in LALP for controlling multiple accesses in a single memory by using FSMs

related to memory access. Multiple memories are handled by different FSMs. Such approach differs from the ones used by some HLS tools, like Vivado HLS [4] and LegUp [5], which are FSM oriented but with a centralized control unit, usually obtained from the control flow graph of the C code being compiled.

A second approach relies on a custom memory architecture that allows multiple memory accesses in a single clock cycle. Figure 4.13 illustrates this architecture. In this case, instead of mapping each array to a single BRAM, each arrays is split into 2^N smaller BRAMs, where N is a parameter of the compiler. Data are indexed through a hash function and each BRAM has its own Memory Management Unit (MMU) for controlling data access. Thus, data from addresses of different BRAMs can be accessed at the same time. If more than one address belongs to one single memory, the corresponding MMU will handle the access to deal with one of them per clock cycle. Depending on the addresses, the data can be accessed out of order, thus, a data binder associates the produced data to their respective datapaths. In terms of syncing, each counter step is delayed according to the number of memory accesses related the MMU which handles most accesses per loop iteration, therefore, it may vary not only according to the number of address but also according to the numbers of BRAMs. A component for controlling the clock step will enable a new iteration once all the data operations are complete.

4.5 Experimental Evaluation

LALP typically focus on a loop pipelining approach and has achieved good results in comparison to ROCCC [6] and C2Verilog [7], as showed in [1], where LALP implementations achieved on average a speedup of 5.60× over C2Verilog [8] for

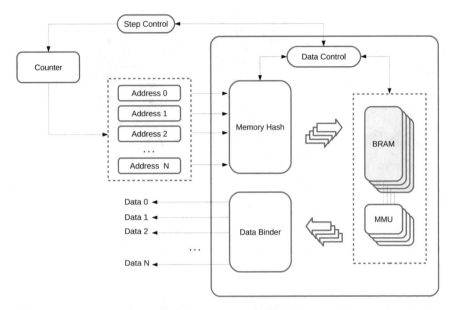

Fig. 4.13 Architecture used in LALP for controlling multiple accesses by using split memory and MMUs

a set of ten kernels, and 1.14× over ROCCC [6] for a set of three kernels. LALP was also compared to two HLS synthesis tools: Vivado HLS 2013.2 [4] and LegUp 3.0 [5]. In such comparison LALP obtained geometric mean speedups of 2.66× and 1.76× over Vivado HLS and LegUp, respectively, for a set of thirteen representative benchmarks [9].

We now present an experimental evaluation of LALP for a set of benchmarks with multiple accesses per clock cycle. In Sect. 4.5.1 we present the results for different versions of a Sobel image processing operator with LALP, while in Sect. 4.5.2 the experiments are focused on evaluate the new components used in LALP architectures, i.e., the proposed custom memory architectures.

4.5.1 The Sobel Operator with LALP

This section presents the experimental results regarding the case of a Sobel operator for four different LALP approaches: the default version, with manual scheduling for memory access (*A*); the alternative version with a FSM-controlled scheduling and using a single BRAM (*B*); the alternative version with 4 MMUs controlling the data accesses (*C*); and an implementation with data reuse (*D*).

The LALP implementations were compared to a number of Sobel designs with different configurations generated by LegUp HLS tool [5] from an equivalent C code. All the experiments were performed using an FPGA from Altera's StratixIV®

family (device EP4SGX530KH40C2). Table 4.3 presents the results regarding LALP implementations, with speedups achieved over the best LegUp design. The results for other LegUp configurations are shown in Table 4.4.

In terms of execution time the LALP case D shows the best result among all designs (i.e., LALP and LegUp). It achieved a speedup of 2.82× over the best LegUp case (config. 1, see Table 4.4). This mainly occurs due to the maximum frequency achieved of 517.9 MHz in addition to the lowest latency. In case D, data reuse is an optimization manually implemented by the programmer in order to reduce the number of operations involving memory accesses.

Considering only the LALP cases without data reuse, the best performance was achieved by the one using 4 BRAMs (case C), which also presented the lowest latency and the higher maximum frequency (286.4 MHz) achieved in comparison to the other LALP cases. All the LALP designs presented maximum frequencies greater than the LegUp best design (see Tables 4.3 and 4.4).

Regarding the use of hardware resources, LALP presents lower values than the LegUp cases. Case B uses the lower number of ALUTs and registers among the LALP cases. It uses 249 ALUTs and 204 registers while the LegUp $v1$, which is the LegUp design with the best performance, uses 922 ALUTs and 760 registers.

When comparing the hardware generated in B (FSM version) to the generated by A (the default approach), it is possible to notice that some registers and muxes used by the latter to balancing and syncing the pipeline are replaced by fewer logic gates to handle the FSM signals in the first one. This justifies the lower resource usage in the FSM approach. In case C, the high resource usage is due to the additional

Table 4.3 Results for different implementations of Sobel filter in LALP using Altera's StratixIV® family (device EP4SGX530KH40C20)

Version	ALUTs	Registers	Max.Freq. (MHz)	Latency	Exec. Time (μs)	Speedup (over LegUp $v1$)
(A) Sobel	314	223	230.6	629	2.73	0.18
(B) Sobel w/ FSM	249	204	276.4	511	1.85	0.27
(C) Sobel w/ MMU	464	481	286.4	255	0.89	0.57
(D) Sobel w/ reuse	278	492	517.9	93	0.17	2.82

Speedup values represent the performance improvements of LALP designs over LegUp $v1$ (see Table 4.4)

Table 4.4 Results for Sobel filter for different LegUp configurations using Altera's StratixIV® family (device EP4SGX530KH40C20)

Version	ALUTs	Registers	Max.Freq. (MHz)	Latency	Dual port BRAM	Loop Pipeline	Exec. Time (μs)
LegUp $v1$	922	760	214.6	109	No	Yes	0.50
LegUp $v2$	951	897	177.4	370	No	No	2.08
LegUp $v3$	918	763	192.9	106	Yes	Yes	0.55
LegUp $v4$	994	866	179.9	304	Yes	No	1.68

hardware required for the MMUs and other control units while in D the higher number of registers is justified by the use of them to control the synchronization over the datapaths. Despite the hardware resources overhead required for using the MMU, such splitting memory approach has a positive impact on the design.

4.5.2 Architectures with Multiple Accesses

As showed in Sect. 4.4, LALP provides different architectures, i.e., FSM and MMU, for handling cases with multiple accesses to the same array per iteration. In this section we present results related to the impact of such architectures in a set of benchmarks that match this condition. The main intention of these experiments is to show and explore some feasible possibilities for implementing LALP architectures for handle distributed memories.

For such purpose we used 6 versions of LALP, as shown in Table 4.5. The chosen benchmarks used for the experiments are: *sobel, prewitt, conv_3×3, median_3×3* and *perimeter*. Each benchmark fits the condition of having multiple accesses per iteration but also presents specific features related to data access. The exceptions are the cases of *sobel* and *prewitt*, both with similar behavior, as data reading follows the same pattern in both cases. Notice that the version of the *sobel* operator presented in this section uses a larger data size than the cases presented in Sect. 4.5.1.

In versions with data partitioning, each of the various BRAMs is proportionally smaller than the single BRAM of the sequential FSM case. The size of each BRAM is equal to M/N, rounded up, where N is the total of BRAMs and M is the size of the array in code. Despite of that, the increasing of the number of BRAMs and the use of components to handle multiple parallel accesses requires a larger amount of FPGA's resources, due to the imposed overhead to control such components.

The resources usage is shown in Figs. 4.14 and 4.15. It is possible to notice the increasing of the number of ALUTs and registers in all the cases, which occurs according to the number of BRAMs, with the FSM approaches using less resources.

Table 4.6 shows the total of parallel accesses and clock cycles per iteration required for each benchmark in the used LALP architectures. The cases of *prewitt*

Table 4.5 Different LALP architectures to handle multiple memory accesses

Version	Architecture	Type of access
FSM	FSM w/ 1 BRAM, no data partitioning	Sequential
MMU_1	MMU w/ 2 BRAMs, w/ data partitioning	Parallel
MMU_2	MMU w/ 4 BRAMs, w/ data partitioning	Parallel
MMU_3	MMU w/ 8 BRAMs, w/ data partitioning	Parallel
MMU_4	MMU w/ 16 BRAMs, w/ data partitioning	Parallel
MMU_5	MMU w/ 32 BRAMs, w/ data partitioning	Parallel

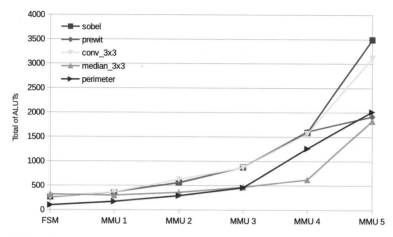

Fig. 4.14 Number of ALUTs used by the different LALP architectures, for an Altera's Stratix IV FPGA, device EP4SGX530KH40C20

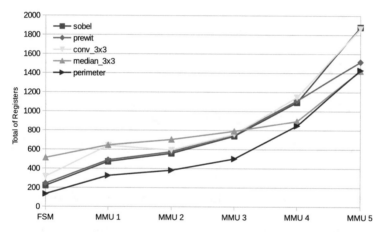

Fig. 4.15 Number of registers used by the different LALP architectures, for an Altera's Stratix IV FPGA, device EP4SGX530KH40C20

and *sobel* require 8 accesses per loop iteration. In both cases, the data are organized as multidimensional arrays and the reading occurs in 3 columns. The *perimeter* algorithm performs 5 accesses, also in 3 columns. However, the data are organized as an unidimensional array. This also occurs in the *conv_3×3* and *median_3×3* cases, with 9 and 3 values loaded per loop iteration, respectively.

As showed in Table 4.6, the data partitioning into various BRAMs may reduce the total of clock cycles per loop iteration required for complete all accesses. As a consequence of that, the total latency of the circuit also decreases, as shown in Fig. 4.16. Such reduction does not depend on how many memories are used, but it is

Table 4.6 Total of parallel accesses and clock cycles per iteration required for each benchmark in the architectures generated from LALP

Benchmarks	Accesses per Iter.	Total of clock cycles					
		FSM	MMU_1	MMU_2	MMU_3	MMU_4	MMU_5
sobel	8	8	6	3	3	3	3
prewit	8	8	6	3	3	3	3
conv_3×3	9	9	7	3	2	1	1
median_3×3	3	3	3	3	3	2	1
perimeter	5	5	4	2	1	1	1

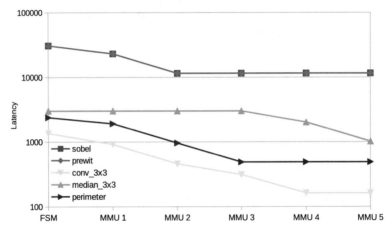

Fig. 4.16 Total latency (# of clock cycles) achieved for each benchmark in the architectures generated from LALP

dependent on the addresses, arrays size and if the data are defined as unidimensional or multidimensional array.

The maximum frequencies achieved in each case are shown in Fig. 4.17. In all cases FSM-based architectures achieved higher frequency values than the MMU-based architectures.

Figure 4.18 shows the speedup of the parallel versions (MMU) over the sequential FSM approach. With respect to the speedups achieved in these experiments the version MMU_5 was never the best option, despite the fact that it shows a higher usage of hardware resources.

The *conv_3×3* case shows the largest difference in when increasing the number of BRAMs, due to subsequent latency reductions. However, the total of hardware resources also increases accordingly. The major speedup occurs in version MMU_4

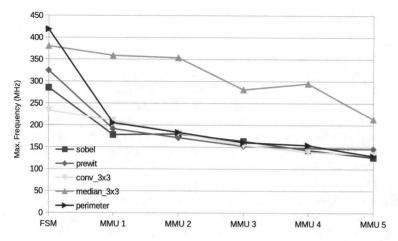

Fig. 4.17 Maximum frequency (MHz) achieved for each benchmark in the architectures generated from LALP

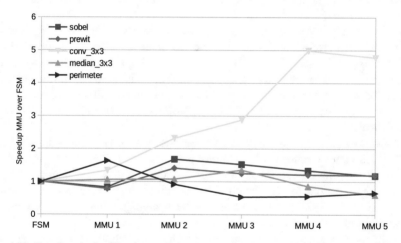

Fig. 4.18 Speedup obtained for the parallel versions (MMU) over the sequential FSM approach

for the *conv_3×3* case. The cases of *sobel* and *prewitt* show the major speedup when using the version MMU_2. The best speedup for the cases *perimeter* and *median_3×3* are the versions MMU_1 and MMU_2, respectively.

As the results show, the use of the proposed custom memory architectures provides the benefit of handling multiple accesses in parallel, with an acceptable cost in terms of hardware requirements. Therefore, such kind of approach can be used to explore alternatives of memory mapping and parallel processing in LALP besides the loop pipelining approach.

4.6   Related Work

The difficulty on using typical hardware description languages such as VHDL and Verilog has lead to a number of approaches that proposed new languages with an increased abstraction level to program reconfigurable hardware. In this section we describe some approaches we think are more related to the LALP language.

Two of the first approaches related to the use of DSLs to produce hardware, the compilers BSAT and StReAm, are presented in [10] and [11]. Such compilers use the PAM-Blox [12] platform, which is a C++ template class library focused on Xilinx XC4000 FPGAs.

BSAT [11] tackles the satisfiability problem in boolean expressions described in Conjunctive Normal Form. The circuits are described in C++ following the PAM-Blox templates while the expressions are mapped into an array of FSMs by using Verilog. The BSAT architecture computes the result of boolean expressions based on partial variable assignments. This is done by using a deduction logic block. The architecture also includes a global controller to start the computation and handle I/O.

StReAm [10] is focused on designing stream objects in C++. It relies on some object-oriented programming features, such as inheritance and operator overloading. The expressions in StReAm are used for generating a dataflow graph where the nodes correspond to the arithmetic operation units. The graph is traversed by the scheduler for calculating the data dependencies and the overall architecture latency. StReAm also creates distributed FIFOs and signals for controlling the sequential components.

As an evolution of the ideas used in the StReAm, in [13] the authors present ASC (A Stream Compiler), a C++ library based on PAM-Blox II, a newer version of PAM-Blox with support for several Xilinx FPGA families. ASC library provides pre-defined functions for handling low-level features, such as registers, signals and clock. Thus, the ASC code includes low-level configurations, such as memory mapping, and features for supporting DSE (Design Space Exploration). The loops ins ASC code are implemented as streams which can be optimized regarding latency, throughput or area.

A scripting interface for designing FPGA circuits, named PyHDL (Python Hardware Description Language), is presented in [14]. PyHDL goal is to reduce the time required for the development of high-level designs with a Python scripting approach. Such approach consists on design circuits with scripts and pre-compiled components, using PAM-Blox templates, and then convert and compile these scripts into new components. Therefore, the benefit is to reduce the compilation overhead and decrease the design time. Thus, PyHDL facilitates a rapid prototyping process while produces a final hardware similar to PAM-Blox.

The MaxCompiler [15, 16] uses MaxJ [15], a Java-like DSL to program and optimize dataflow engines (DFEs). The DFEs are implemented in Maxeler Technologies[2] boards and consist of large memory banks (DRAMs) with programmable elements to process the computation. The MaxJ uses an approach where the architecture is composed by CPU and kernel accelerators. The kernels implement the computationally intensive parts of the algorithms, mainly consisting in loops. A data compression approach is used to store data into on-chip memory banks, where the data are stored as tiles and the arrays layouts are automatically computed by the compiler in order to optimize the data streaming.

OpenSPL (Open Spatial Programming Language) [17] is a language also based on DFEs. It implements the concept of spatial computing, where the program's data flow and control flow are decoupled, with the operations running in parallel by default. OpenSPL architectures rely on Spatial Computing Substrates (SCS), which are dataflow units scheduled to optimize the data accesses and minimize the data flow. A SCS may have one or more parallel kernels, interconnected by a common dataflow. It also includes arithmetic units, customizable interconnect and memories.

In [18, 19] the authors present an approach named FCUDA, where RTL (Register Transfer Level) code for FPGAs can be generated based on code written for GPUs with the NVIDIA's CUDA API. For such, the FCUDA programmer is required to modify the CUDA kernels by inserting proper *#pragma* directives in order to explicit the parallelism of the kernels. FCUDA applies code transformations to the input code by using the Cetus Compiler [20]. In order to produce the output hardware corresponding to the concurrent kernels, FCUDA uses the Autopilot [21] HLS tool to perform the high-level synthesis of the transformed code. Other features, such as granularity, memory organization and synchronization of the kernels, can also be specified by using the *#pragmas* provided in FCUDA.

Haydn-C [22] is a DSL where the programmer is allowed to include pipelining directives in the code. It combines cycle-accurate and behavioral design methodologies. In such approach the programmer can write a behavioral description of the code but can also insert directives to control the design scheduling. The Haydn-C approach uses two types of timing semantics with strict and flexible timing models. The *strict* timing model defines a set of rules related to sequential/parallel execution flow and duration of the operations. The *flexible* model is meant for dealing with code transformations by relaxing some of the *strict* rules, as long as the process result remains correct, in order to make transformations feasible. Haydn-C is a component-based language based on Handel-C and, therefore, uses *par* instructions to define parallel sections.

Single Assignment C (SA-C) [23], is a C-like DSL used for developing FPGA-based architectures with a co-design approach and with the single assignment restriction. SA-C compiler allows the programmer to directly insert VHDL code, since it performs the automatic connection of the hardware blocks. As a co-design approach, the sequential pieces of code are translated into C code and run on a CPU. For hardware generation the compiler uses a Data Dependency Control Flow graph computed from the input code. An architectural graph is generated and includes

---

timing information used for producing the final VHDL. The compiler defines which loops will be executed in the FPGA. Moreover, a loop in SA-C is handled in such way that it returns the result of the computation at the end of the iterations.

Handel-C [24] is a hardware imperative programming language where FPGA components, such as memories and FIFOs, are exposed to the programmers. In Handel-C, each assignment statement is assumed to have one clock cycle of latency and the concurrency is explicitly specified as code sections using the *par* instruction. The programmer uses decomposition of expressions to control the operations assigned to the same cycle in an expression.

SystemC [25, 26] is a standard C++ class library for system and hardware design where concurrent processes (C++ functions) can be defined within a module hierarchy, with communication channels used for inter-module/process communication. Once the modules are created, a simulation step is performed in order to schedule and synchronize the processes. A set of events (such as reading and writing) is used by the scheduler for controlling the execution order of processes, including those that are supposed to execute concurrently.

CHiMPS [27] is a HLS compiler for generating FPGA accelerators from a generic ANSI-C code. It uses a spatial dataflow hardware model where each node in a program's dataflow graph is instantiated as a physical node in the resulting FPGA accelerator. CHiMPS runs on a hybrid CPU-FPGA using the MicroBlaze [28] soft core processor and relies on a many-cache memory model with many small distributed caches to support simultaneous memory requests.

Recently, the major FPGA manufacturers have offered compilers and runtime libraries for OpenCL [29], as showed in [30, 31], which is a programming standard for general-purpose heterogeneous computing. SOpenCL (Silicon OpenCL) [32] is an OpenCL-based approach for developing hardware accelerators. It uses source-to-source transformations to produce parallel kernels from the input program. HDL code for such kernels is produced to fit an architectural template where data accesses and computations are decoupled in order to reduce latency effects. Although it is a promising initiative, it is focused on suitable OpenCL computations and may not be adequate for compiling many computations.

LALP distinguishes from the previous research efforts as it is a special purpose language focused on to program FPGA-based accelerators for loops. In LALP, programmers specify the accelerators in a behavioral level instead of their architectures. Differently from the other efforts, LALP compiler assumes that all the statements are concurrent and what makes them sequential are the control/data dependences between them and the explicit use of a scheduling qualifier '@'.

4.7 Conclusions

This chapter presented a new version of LALP, a special-purpose DSL to program FPGA-based hardware accelerators. The LALP approach has as goal to provide a language that covers a wide range of possibilities for designing FPGA-based accelerators. Considering the presented LALP features, such as abstraction level (C-like syntax) and custom architecture, it is expected that with LALP developers are able to tackle issues that are not currently handled by other FPGA development approaches, like the use of HLS tools, and/or when these approaches can not achieve sufficient performance.

The LALP based development efforts are lower than the ones to implement designs using HDLs but it requires more efforts than the use of a HLS tool, specially when the programmer's intervention is necessary to manage the loop scheduling in cases of multiple memory accesses as the automatic achieved scheduling does not always lead to the best design. To deal with such level of complexity, LALP provides different options for handling memory accesses scheduling. It includes an alternative syntax where the programmer can avoid the explicit insertion of delays. This version relies on specific memory architectures. Currently the LALP compiler only supports one-way accesses for each array, i.e., reading or writing operations but not both.

The use of multiple BRAMs enables LALP developers to implement loop transformations related to distributed memories. Nevertheless, developers still needs to take into account the number of addresses and arrays and also how the data are partitioned when defining the total of BRAMs, once both these parameters impact the number of cycles per iteration and the data throughput.

The results achieved by using LALP motivate us to continue improving the language and the compiler. Ongoing work focus on including more advanced compiler support to suggest locations of LALP '@' qualifiers and their values and also improved ways to handle the memory accesses scheduling, by including components such as FIFOs. As LALP also includes specific pipelined modules for floating-point operations, future work will extend the support to floating-point operations and include the support to custom floating point data types in these modules.

Acknowledgements The authors would like to thank FAPESP (the Foundation to Support Research of the State of São Paulo) for the financial support provided.

References

1. Menotti, R., Cardoso, J.M.P., Fernandes, M.M., Marques, E: LALP: a language to program custom FPGA-based acceleration engines. Int. J. Parallel Prog. **40**(3), 262–289 (2012)
2. Cardoso, J.M.P., Neto, H.C.: Macro-based hardware compilation of Java^TM bytecodes into a dynamic reconfigurable computing system. In: Proceedings of the 7th Annual IEEE Symposium on Field-Programmable Custom Computing Machines, pp. 2–11. IEEE (1999)
3. IEEE: IEEE Standard for Binary Floating-Point Arithmetic, ANSI/IEEE Std 754-1985 (1985)
4. Feist, T.: Vivado Design Suite. White Paper, Xilinx Inc. (2012)
5. Andrew, C., et al.: LegUp: an open-source high-level synthesis tool for FPGA-based processor/accelerator systems. ACM Trans. Embed. Comput. Syst. **13**(2), 24:1–24:27 (2013)
6. Villarreal, J., Park, A., Najjar, W., Halstead, R.: Designing modular hardware accelerators in c with roccc 2.0. In: 18th IEEE Annual International Symposium on Field-Programmable Custom Computing Machines, pp. 127–134. IEEE (2010)

7. Huong, G.N.T., Kim, S.W.: Gcc2verilog compiler toolset for complete translation of c programming language into verilog hdl. ETRI J. **33**(5), 731–740 (2011)
8. Rotem, N.: Online: http://www.c-to-verilog.com/ (2010)
9. de Oliveira, C.B., Cardoso, J.M.P., Marques, E.: High-level synthesis from C vs. a DSL-based approach. In: IEEE International Parallel Distributed Processing Symposium Workshops (IPDPSW), pp. 257–262 (2014)
10. Mencer, O., Hubert, H., Morf, M., Flynn, M.: StReAm: object-oriented programming of stream architectures using PAM-Blox. In: IEEE Symposium on Field-Programmable Custom Computing Machines, pp. 309–310 (2000)
11. Mencer, O., Platzner, M., Morf, M., Flynn, M.: Object-oriented domain specific compilers for programming FPGAs. IEEE Trans. Very Large Scale Integr. VLSI Syst. **9**(1), 205–210 (2001)
12. Mencer, O., Morf, M., Flynn, M.: PAM-Blox: high performance FPGA design for adaptive computing. In: IEEE Symposium on FPGAs for Custom Computing Machines, pp. 167–174 (1998)
13. Mencer, O.: ASC: a stream compiler for computing with FPGAs. IEEE Trans. Comput. Aided Des. Integr. Circuits Syst. **25**(9), 1603–1617 (2006)
14. Haglund, P., Mencer, O., Luk, W., Tai, B.: Hardware design with a scripting language. In: Peter, Y.K.C., Constantinides, G.A. (eds.) Field Programmable Logic and Application. Lecture Notes in Computer Science, vol. 2778, pp. 1040–1043. Springer, Berlin/Heidelberg (2003)
15. Fu, H., Gan, L., Clapp, R.G., Ruan, H., Pell, O., Mencer, O., Flynn, M., Huang, X., Yang, G.: Scaling reverse time migration performance through reconfigurable dataflow engines. IEEE Micro **34**(1), 30–40 (2014)
16. Pell, O., Bower, J., Dimond, R., Mencer, O., Flynn, M.J.: Finite-difference wave propagation modeling on special-purpose dataflow machines. IEEE Trans. Parallel Distrib. Syst. **24**(5), 906–915 (2013)
17. The OpenSPL Consortium: OpenSPL: Revealing the Power of Spatial Computing. White Paper, The OpenSPL Consortium (2013)
18. Papakonstantinou, A., Gururaj, K., Stratton, J.A., Chen, D., Cong, J., Hwu, W.-M.W.: Efficient compilation of cuda kernels for high-performance computing on FPGAs. ACM Trans. Embed. Comput. Syst. **13**(2), 25:1–25:26 (2013)
19. Papakonstantinou, A., Gururaj, K., Stratton, J.A., Chen, D., Cong, J., Hwu, W.-M.W.: High-performance cuda kernel execution on FPGAs. In: Proceedings of the 23rd International Conference on Supercomputing (ICS), pp. 515–516. ACM, New York (2009)
20. Dave, C., Bae, H., Min, S.J., Lee, S., Eigenmann, R., Midkiff, S.: Cetus: a source-to-source compiler infrastructure for multicores. Computer **42**(12), 36–42 (2009)
21. Zhang, Z., Fan, Y., Jiang, W., Han, G., Yang, C., Cong, J.: AutoPilot: a platform-based ESL synthesis system. In: Coussy, P., Morawiec, A. (eds.) High-Level Synthesis, pp. 99–112. Springer Netherlands, New York (2008)
22. Coutinho, J.G.F., Jiang, J., Luk, W.: Interleaving behavioral and cycle-accurate descriptions for reconfigurable hardware compilation. In: 13th Annual IEEE Symposium on Field-Programmable Custom Computing Machines, pp. 245–254 (2005)
23. Najjar, W.A., Bohm, W., Draper, B.A., Hammes, J., Rinker, R., Beveridge, J.R., Chawathe, M., Ross, C.: High-level language abstraction for reconfigurable computing. Computer **36**(8), 63–69 (2003)
24. Agility Design Solutions Inc.: Handel-C Language Reference Manual. White Paper (2007)
25. IEEE Comp. Society: IEEE Standard for Standard SystemC Language Reference Manual. IEEE Std 1666-2011 (2012)
26. Grötker, T., Liao, S., Martin, G., Swan, S.: System design with SystemC. Springer, New York (2002)
27. Putnam, A., Bennett, D., Dellinger, E., Mason, J., Sundararajan, P., Eggers, S.: CHiMPS: a C-level compilation flow for hybrid CPU-FPGA architectures. In: International Conference on Field Programmable Logic and Applications, pp. 173–178 (2008)
28. Xilinx: MicroBlaze Processor Reference Guide (2014)
29. Khronos OpenCL Working Group: The OpenCL Specification, Version 2.0 (2014)

30. Altera Corp.: Altera SDK for OpenCL: Getting Started Guide (2015)
31. Wirbel, L.: Xilinx SDAccel: a unified development environment for tomorrow's data center. Technical Report, The Linley Group Inc. (2014)
32. Owaida, M., Bellas, N., Daloukas, K., Antonopoulos, C.D.: Synthesis of platform architectures from OpenCL programs. In: IEEE 19th Annual International Symposium on Field-Programmable Custom Computing Machines (FCCM 2001), pp. 186–193 (2011)

Part III
Clocks and Temporal Issues

Chapter 5
Enabler-Based Synchronizer Model for Clock Domain Crossing Static Verification

M. Kebaili, K. Morin-Allory, J.C. Brignone, and D. Borrione

5.1 Introduction

The context of this research is the design of large circuits, assembling IP's and blocks coming from various design teams, each with their own clock. As an example, circuits for games or set top box applications easily include in excess of 25 blocks. To guarantee low power consumption, the clock of a block is set at the lowest frequency compatible with the block needed execution time. Thus, all the clocks have their own speed, and there is no guarantee that clock cycles are multiple of a basic master clock, nor that two clocks are phased.

Each block constitutes a clock domain, and any two clock domains are considered mutually asynchronous: the design is globally asynchronous locally synchronous (GALS). Clock domain crossing (CDC) is the transmission of information between two clock domains. A synchronizing module (called *synchronizer* hereafter) is needed to connect a source signal, output of a flip-flop in a transmitter domain, to the input of a destination flip-flop in a receiver domain, because the sampling by the receiver clock may happen before the input signal has reached its correct stable value. It is therefore essential to guarantee the correctness of the communication protocol and the synchronization between the modules [4].

One technique consists in implementing monitors in the design, which perform online checking and correction during the circuit operation [10, 11]. The drawback

M. Kebaili • J.-C. Brignone
STMicroelectronics, Paris and Grenoble, France
e-mail: mejid.kebaili@st.com; jean-christophe.brignone@st.com

K. Morin-Allory • D. Borrione (✉)
TIMA Laboratory, University of Grenoble-Alpes, Grenoble, France
e-mail: katell.morin-allory@imag.fr; dominique.borrione@imag.fr

© Springer International Publishing Switzerland 2016
R. Drechsler, R. Wille (eds.), *Languages, Design Methods, and Tools
for Electronic System Design*, Lecture Notes in Electrical Engineering 385,
DOI 10.1007/978-3-319-31723-6_5

is the significant overhead on the circuit size. To avoid online monitors, several synchronization methods have been proposed and proved correct [12].

Typical academic work [1] considers the presence of a given (new) synchronizer between clock domains. What needs to be verified is correctness properties over the protocol between the transmitter and the synchronizer on the one hand, the synchronizer and the receiver on the other hand.

In contrast, in an industrial size design, synchronizers are not so well identified. This is due to the fact that design teams and verification teams are distinct. And designers often develop their own synchronizers for performance reasons. This problem is so crucial that several industrial tools [2, 14, 16] have been developed to help automate the verification of CDC communications. Such tools rely on (1) libraries of components to identify synchronization structures in the design, and (2) constraints specified by the verification engineer.

If the synchronizers are correctly recognized, the correctness verification can be attempted. Unfortunately, some structures are only partially recognized; in this case, the tools demand a strong interaction with the verification engineer to provide more constraints, case by case. The drawback of this method is that it is not reusable. Designs evolve over time, and their verification must be repeated. It is thus important to automate the recognition of synchronization structures and limit the manual addition of constraints.

The original contribution of our work is a new synchronization structure model and its associated properties. This model is reusable, and limits the need for user interaction.

This paper first reviews the metastability problem, the classical synchronizers as countermeasures and the state of the art verification methods that assume a proper identification of the synchronizers. Then Sect. 5.3 explains the reasons for and the consequences of a partial recognition of the synchronizers. Section 5.4 discloses our strategy for synchronizer identification, and its associated properties to be verified on the design. Section 5.5 presents the performance results of our method, and Sect. 5.6 gives our conclusions.

5.2 CDC Verification State of the Art

5.2.1 Metastability: A CDC Major Issue

To operate correctly, the sequential elements that ensure the synchronization of input data must be loaded with an input signal that remains stable during the clock's setup and hold times. When one of these two requirements is violated, the state of the output signal is not predictable. This phenomenon is called *metastability*. During a signal transition from a source clock domain, its capture by the destination may cause the generation of a metastable state at the output indefinitely. Figure 5.1 shows a violation of a destination clock's setup and hold times, generating a metastable

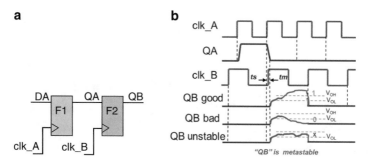

Fig. 5.1 Metastability. (**a**) Clock domain crossing (CDC). (**b**) Metastable state generation

state at the output *QB* of the sequential element *F2*. We can see that the metastable state resolution of QB can be done at the good state, the bad state or remain unstable.

The random nature of metastability can cause errors [3] in case of a data bus transfer. Metastable bus bit signals can be sampled at different states in the destination domain and result in incoherent data propagation.

Metastability is not the only issue related to clock domain crossing. Other killer problems can appear such as glitches capture, incoherent data propagation, or data loss [5, 17]. In this chapter, we focus on data synchronization issues.

5.2.2   CDC Design Methods

The metastability problem cannot be eliminated, this is why a countermeasure is needed to limit its effects [7, 13]. The principle consists in adding a latency, via the insertion of sequential elements, in order to ensure the resolution of the metastability before propagating the signal through the system. It is possible to estimate the time required to return to stability, and therefore the number of registers to be added (depth), by calculating the "Mean Time Between Failure" (MTBF). Simulation-based fault injection is the usual technique to verify that metastability does not propagate. In contrast, little is published on the formal verification of stability at the analog level [15].

The principle of cascaded flip-flops after the capture of the signal in the destination domain is called a "multi-flop synchronizer". Figure 5.2 shows a good resolution of metastability and stable signal transfer (QS) after synchronization.

The multi-flop synchronizer is used only to avoid X propagation due to metastability generation. However, data loss can appear in the case of a bad resolution of metastable states. This second issue can be reduced to a delay holding the transferred data during multiple destination clock cycles. However, in the case of a data bus transfer (that is not Gray encoded) through a CDC, incoherent data can be propagated through multi-flop synchronizers. Figure 5.3 shows a data bus transfer through a CDC generating incoherent data at the multi-flop synchronizer output.

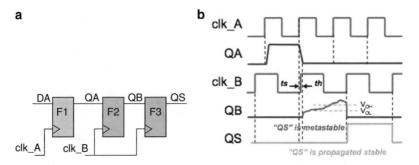

Fig. 5.2 Multi-flop synchronizer. (**a**) CDC with multi-flop. (**b**) Good resolution of the metastability

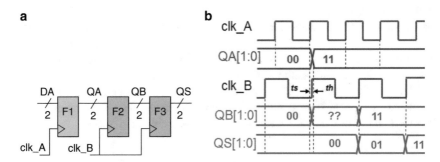

Fig. 5.3 CDC of data bus with multi-flop. (**a**) Multi-flop synchronizer. (**b**) Bad resolution of the metastability

In order to ensure both data coherency and metastability resolution, control-based synchronizers are used [7, 8]. The idea is to use an additional control signal, crossing through the same clock domains and synchronized by multi-flop, to enable the capture of a data bus in the destination domain.

The data signals through the CDC must be established and remain stable before the arrival of the control signal. Data are thus captured in a stable condition and at the same clock cycle in the destination domain.

There are many structures of synchronization by a control signal. The most common are the recirculation multiplexer, the AND gate and the Enable-based schemes.

Figure 5.4 shows an Enable-based scheme with a control signal CDCcontrol coming from the source clock domain, going through a multi-flop synchronizer and enabling the CDCdata capture at DstEN.

In addition to data stability and coherency, a correct communication must guarantee that no new data is sent from the source domain as long as the destination domain has not captured the signals being transferred.

The solution consists in the addition of a feedback control signal that ensures the exclusivity between the send and capture operations. This signal enables the

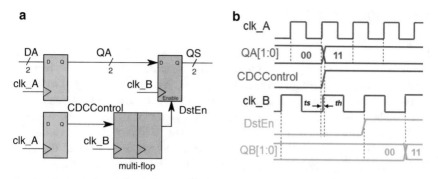

Fig. 5.4 CDC of data bus with control-based synchronizer. (**a**) Control-based synchronizer. (**b**) Good data bus transfer

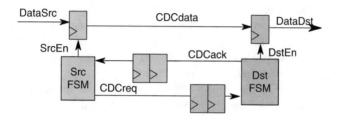

Fig. 5.5 Handshake synchronizer

sending of data on the bus: it comes from the destination clock domain, and goes to the source one through a multi-flop synchronizer. This type of structure is called *protocol-based synchronizer*, and includes the handshake and FIFO (First In First Out) schemes [7, 8].

Figure 5.5 shows a handshake synchronizer. The source domain emits data on the CDCdata bus, together with the CDCreq control signal. CDCreq goes through a multi-flop synchronizer, which guarantees that it will trigger the loading of the data bus in the destination domain after CDCdata has reached stability. The synchronized control signal may be directly used as "enable", as was shown in Fig. 5.4. In Fig. 5.5, we illustrate the presence of an additional finite state machine (FSM) that drives the DstEn enable signal in the destination domain. The FSM also generates the CDCack acknowledgement signal, i.e. the reply of the destination to the source domain. Again, CDCack goes through a multi-flop synchronizer before reaching the source control FSM that generates the successive loading of source data at SrcEn.

In the sequel, we shall talk of a *control synchronizer* for multi-flop, *partial data synchronizer* for control-based schemes, and *full data synchronizer* in case of protocol-based structures. In summary, the data synchronizers (partial and full) are structures allowing a transfer of multiple signals between two clock domains without metastability generation.

5.2.3 CDC Static Verification

To avoid potential CDC issues in the designs, major EDA providers (Cadence, Mentor Graphics, Synopsys, . . .) propose CDC static checkers to perform two kinds of verification [9], according to the time available, the required quality of results and the complexity of the design: structural verification and functional verification.

The CDC *structural verification*, consists in a pattern matching between an input hardware description (at RTL or Gate level) and a set of library cells describing CDC structures. In case of match, the parts of the design corresponding to the library descriptions are reported through instrumented models. However, at this step, the tool analyses only the interconnections between elements and not the functionality; thus, even if a structure is identified, it must be functionally verified.

The CDC *functional verification* calls upon a model-checking engine to prove formal properties on an architectural model identified during the structural verification. The properties are linked to the library cells used during pattern matching. After the functional verification, the results are split into two conclusions: proven or failed. At this step we can assume that a property proof ensures a design correctness and that a failure requires a hardware description debugging.

Figure 5.6 presents the ideal CDC verification flow implemented in CDC checkers. Structural verification produces a model of the design in which the CDC structures have been detected and classified; this model, together with the appropriate formal properties, is verified by model checking. Obviously, the structural verification results depend on the pattern libraries coded into the EDA tool, and the functional verification is based on structural results.

CDC analysis covers multiple functional issues including glitch capture, incoherent data propagation, and data loss. The synchronizer detection and verification are pre-requisite for an exhaustive CDC analysis, and are thus the focus of our work.

In this context, the CDC *structural verification* consists in the data synchronizer detection. In most CDC checkers, those structures are split into three types: Handshake, FIFO and control-based. Also, The CDC *functional verification* is the analysis of the synchronization protocol to guarantee the correctness of the data transfer through the CDC. In most industrial CDC tools, the properties depend on the synchronization structure detected: request/acknowledge protocol for Handshake, overflow/underflow for FIFO and data stability for control-based synchronizers.

Figure 5.7 presents the ideal verification flow for data synchronizers implemented in CDC checkers.

5.3 CDC Verification on Industrial Designs

In most industrial designs, the data synchronizers identification presents limitations that can be grouped in two families: (1) the false issues, which leaves some

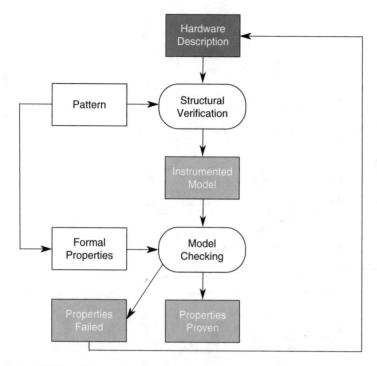

Fig. 5.6 Ideal CDC verification flow

synchronization structures undetected and (2) the false information, which identifies false synchronization structures. The false issues family depends on the permissiveness of the structural verification engine, and more precisely the pattern matching conditions coded in the libraries.

5.3.1 Structural Detection Approach

To illustrate the above mentioned issues, consider the FIFO synchronizer detection. Most industrial checkers are based on the identification of three groups of signals: the data bus, the write address bus and the read address bus. A match is found if all the three following conditions are met:

- the data bus is not synchronized by multi-flop, while the read and write address busses are;
- the write address bus comes from the source clock domain and goes both to the memory data registers and to the destination clock domain;
- the read address bus comes from the destination clock domain and goes both to the memory reading logic, and to the source clock domain.

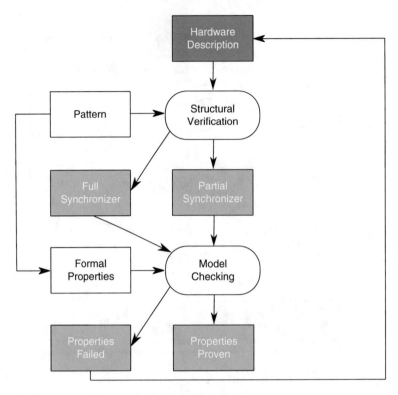

Fig. 5.7 Ideal data synchronizers verification flow

Figure 5.8 shows such a FIFO synchronization structure (of depth 4). Let *ds* be the
size of the source data bus *DataSrc*. The data register file of the FIFO belongs to the
source domain as well as its output bus of size $4 \times ds$. In the destination domain,
there is only one data register. *WrAdd* is the write address bus going from the control
part *WrFSM* of the data registers to the MFDst destination multi-flop synchronizer.
The 4-bit signal *WrEn*, which selects the FIFO element to be written, is the one hot
encoding of *WrAdd*. RdAdd is the read address bus going both to the destination
reading logic multiplexer and to the MFSrc source multi-flop synchronizer. *RdEn* is
the control signal enabling the data capture in the destination register. WrAdd and
RdAdd are Gray encoded. In the example of Fig. 5.8, they are two bits each.

The source WrFSM is a small automaton that compares the synchronized address
of the data currently read with the address of the previously written data in the FIFO,
to prevent writing over a data not yet read. The destination RdFSM is a distinct
automaton that compares the synchronized address of the data being written to the
address of the data currently read, to avoid reading twice the same information when
the FIFO has emptied.

Let RdAdd be the address of the next data to be read, PrWrAdd and WrAdd
the addresses of the previous and next data to be written (all three registers are

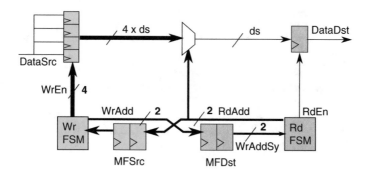

Fig. 5.8 FIFO synchronizer model

reset to 0), and "+" be the modulo addition. The correct behavior of the FIFO automata is specified by these safety properties, which are written in PSL [6] at an abstract level for readability (the Gray encoding is not taken into account):

```
// no write in a full FIFO
assert always WrEn -> RdAdd != PrWrAdd + 1;
// no read from an empty FIFO
assert always RdEn -> RdAdd != WrAdd;
```

It is expected that these properties have already been proved by model checking, assuming that the source and destination have the same clock. The CDC functional verification aims at proving that this behavior is maintained across two distinct clock domains.

However, a trivial issue is a FIFO integrating a sequencer on the reading logic. In this case, the multi-bit read address bus is split into sub-buses, allowing a full-speed data transfer, before going to the reading logic. This violates the condition for the read address bus identification: it is not the same bus signals that go to the memory reading logic multiplexer and to the source domain multi-flop synchronizer, thus the data synchronizer is not detected. This is illustrated in Fig. 5.9, where the RdAdd read address bus is input to a FSM generating two address pointers sub-buses before going to the multiplexer of the reading logic. In the case of Handshake structures, the same type of issue appears if the design to be verified contains a structural description absent from the pattern libraries.

To avoid these buggy results without building design dependent library cells, some industrial tools allow the partial detection of data synchronizers through a control-based synchronizer model [14, 16]. The idea is to detect the signal that controls data stability after CDC. The matching conditions are a signal coming from the source clock domain, going to the destination domain through a multi-flop synchronization and being captured by the data destination register. For a FIFO, only the data bus and part of the write address pointers bus conditions are required to detect a data synchronizer (the write address is not required to go to the source data registers). The increased pattern matching permissiveness can generate the detection of false synchronizer structures; this is the explanation for the

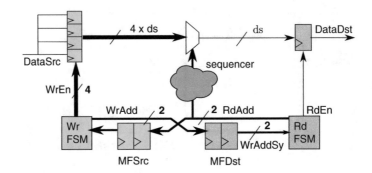

Fig. 5.9 FIFO synchronizer detection issue in case of ping-pong architecture

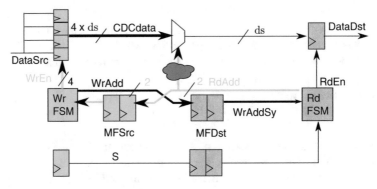

Fig. 5.10 Issues related to data synchronizers partial detection

false information issue mentioned previously. In this case, some design constraints must be added manually by the verification engineer, to help the partial detection of data synchronizers by the tool. Figure 5.10 presents a FIFO structure bad partial detection, with the identification of S instead of WrAdd as control signal.

5.3.2 Functional Verification Approach

Since the functional verification is done on the model identified during the structural verification (see Fig. 5.6), the ideal CDC flow presents some limitations when the data synchronizers are partially identified: the extracted model is not complete and corresponds to the control based synchronizer of Fig. 5.4. So the properties to be verified relate only to the stability of the data just written, data coherency is left out.

The main property is evaluated at each rising edge (function *posedge* in PSL) of both the source clock *ClkSrc* and the destination clock *ClkDst*. It states that, whenever the read enable signal *RdEn* is active, the *CDCdata* signal must be stable. In PSL:

```
default clock = (posedge clkSrc) | (posedge clkDst);
assert always (RdEn -> stable (CDCdata));
```

The property related to control based synchronizers verification does not distinguish the synchronizer data input and the control signal enabling the data transfer through the CDC. In the case of hierarchical verification, a stand-alone analysis of the data synchronizer can be performed. But in an industrial design, the data synchronizers between two power domains are not always wrapped into a module. As a consequence, when verifying a flattened design, the formal engine analyzes the whole cone of influence of the property, i.e. all the sequential and combinational sub-circuits that influence, in one or more cycles, the value of the property operands. So both the source register bank control path and the data path are involved. Since the data synchronizer boundaries are not properly identified, the analysis may go back to the source clock domain logic, resulting in state explosion and inconclusive results.

Figure 5.9 illustrates this problem with a property on the data in the case of partial FIFO detection. With respect to the previously written property, RdEn has to be "anded" with the synchronized address of the data that has been written. At high level, the property says that, for all addresses i, ($0 \leq i \leq 3$), if the read enable signal *RdEn* is active and the synchronized write address signal *WrAddSy* contains value i, then *CDCdata[i]* must be stable.

```
default clock = (posedge clkSrc) | (posedge clkDst);
%for i in 0..3 do
    assert always
        (RdEn && (WrAddSy==%{i}) -> stable CDCdata%{i});
```

In reality, taking into account the address coding, the property has to be written for each address value:

```
assert always
    (!WrAddSy[0] && !WrAddSy[1] && RdEn) -> stable(Data0))
assert always
    ( WrAddSy[0] && !WrAddSy[1] && RdEn) -> stable(Data1))
assert always
    ( WrAddSy[0] &&  WrAddSy[1] && RdEn) -> stable(Data2))
assert always
    (!WrAddSy[0] &&  WrAddSy[1] && RdEn) -> stable(Data3))
```

The formal analysis of the property may structurally incorporate all the circuit elements that have a path leading to the property operands (fan-in cone), possibly up to the primary inputs. The complexity of the model to be verified thus far exceeds the complexity of the synchronization structure alone.

In order to perform CDC functional verification with conclusive results, cut points declarations under the form of constraints must be added manually to give the synchronizer boundaries to the tool. Figure 5.11 presents the STMicroelectronics CDC verification flow for data synchronizers implemented in CPU based designs. It shows the constraints specification requirement for the structural identification (synchronizer partial model detection) and the functional verification (model boundaries definition).

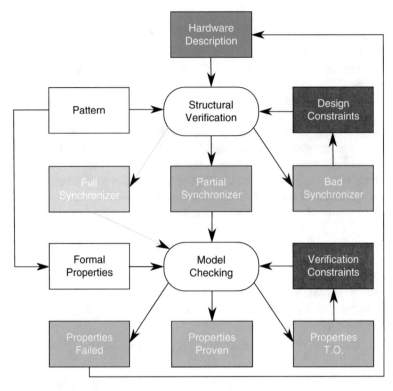

Fig. 5.11 Constraint-based data synchronizers verification flow

Unfortunately, this flow is too human work expensive to meet the time to market requirements. The design and verification constraints definition is tool dependent, design dependent and requires a lot of CDC knowledge. This motivates the elaboration of a new design synchronizer model called "enabler-based synchronizer" with a set of properties for synchronization protocol verification: our aim is to perform CDC static verification on non standard data synchronizers without the need for additional constraints definition.

5.4   Enabler-Based Synchronizer Proposal

5.4.1   Data Synchronizer Review

Structurally, a data synchronizer is composed of control signals, allowing the transmission from the source domain and the capture into the destination domain. The synchronization control signals are sampled into both clock domains and synchronized by multi-flop.

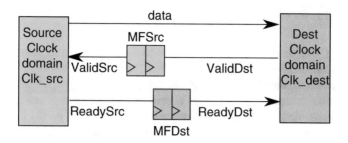

Fig. 5.12 High level data synchronizer overview

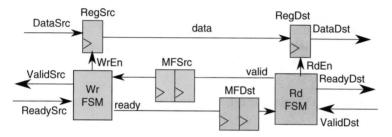

Fig. 5.13 High level data synchronizer architecture

Figure 5.12 presents the synchronized transfer of the data bus between the Source and Dest clock domains. The data synchronizer is composed of two control signals synchronized respectively by Source and Dest multi-flop: valid to enable data sending and Ready to allow the data capture. In the figure, ValidDst (ReadySrc) and ValidSrc (ReadyDst) are the names of the signal before and after resynchronization.

At architectural level, a data synchronizer can be modeled as a block composed of two clock domains, each of which integrates the data registers and a control logic block. The interconnection between the two control logic blocks is composed of one signal going from the source to the destination, to indicate that the source is ready to send, and one signal going from the destination to the source, to indicate that the destination part has received the data. The control signals are synchronized by multi-flop before being captured by the control logic blocks. Also, the control logic blocks have one output to enable the data sampling by the source and destination data registers. These enablers have in their fan-in both control signals in each clock domain, one before the multi-flop synchronizer and one after.

Figure 5.13 shows the full architecture of a data synchronizer, with the data registers RegSrc and RegDst, the two control logic blocks WrFSM and RdFSM, and two enable signals WrEn and RdEn to activate data sending from the source clock domain and data capture into the destination clock domain.

5.4.2 Enabler-Based Synchronizer Model

Considering that the principal issue related to data synchronizer detection is the complete identification of the interconnection between the writing control logic and the reading control logic, we propose a model of data synchronizer based on the four-phase protocol scheme. This approach allows a compromise between a complete recognition of the communication structure, and the partial identification of the synchronizer that is currently available in the CAD tools. This can be easily understood if we take the example of the FIFO. In our model, the conditions allowing FIFO structure recognition are

1. the data crossing path;
2. the control signal coming from the source clock domain and going both to the source data register and to the destination one through a multi-flop synchronizer, which we call the synchronization enabler path (in the fan-in cone of the controller path).

So, the data synchronizer model, the control part of which is detailed in Fig. 5.14, is composed of:

- the data crossing path (ensuring a CDC): it is the $DataSrc \rightarrow Data \rightarrow DataDst$ path;
- the data coherency control path (controller feedback): it is the $RdAdd \rightarrow MFSrc \rightarrow ValidSrc$ path; this control path is not needed for structural recognition, but it is necessary for functional verification;
- the data stability control path (controller synchronized): it is the $WrAdd \rightarrow MFDst \rightarrow WrAddSy \rightarrow ReadyDst$ path.

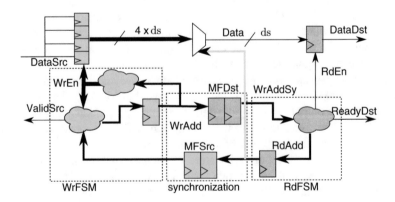

Fig. 5.14 Data synchronizer model

5.4.3 Enabler-Based Protocol Definition

Now that we have developed a model with enough information to guide the structural detection engine, we have to determine the properties to be verified on the data synchronizer. The two essential requirements are:

- the data crossing stability during the capture into the destination domain (no metastability generation),
- the data coherency verification during the crossing (no data loss).

These requirements are fulfilled by the following four-steps synchronization protocol:

1. No data reading when data is being written (stabilization),
2. No data writing when data is being read (capture),
3. No new data writing until data has been captured,
4. No new data reading until new sent data has stabilized.

So the writing and reading phases must be exclusive for each given data (for each address in the case of a FIFO). Figure 5.15 shows the synchronization protocol proposed, which is a handshake.[1]

The synchronization protocol is specified by the following PSL assertions, which are the properties to be formally verified over the corresponding control signals of the extracted enabler-based synchronizer. The first property states that, whenever signal *WrEn* has a falling edge (function *fell* in PSL), then it must remain low as long as *RdEn* is not active. The second property is identical, exchanging the roles of *WrEn* and *RdEn*. The third and fourth property together state that *WrEn* and *RdEn* cannot be active simultaneously.

```
assert always ( fell (WrEN) -> !WrEN until RdEN);
assert always ( fell (RdEN) -> !RdEN until WrEN);
assert always (WrEN -> !RdEN);
assert always (RdEN -> !WrEN);
```

Fig. 5.15 Data synchronization protocol

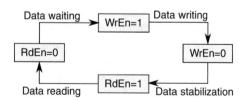

---

[1]Beware that in CDC verifications tools, the handshake primitive structure is a self-acknowledgement. Here we mean a real handshake.

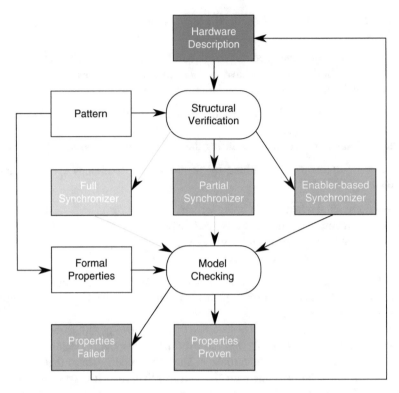

Fig. 5.16 Robust data synchronizers verification flow

It is important to note that these properties only involve the enabling control signals, not the data, which considerably reduces the cone of influence being considered by the model checking engine, with a dramatic reduction on the formal verification execution time.

With our data synchronizer model, the CDC static verification flow is close to the ideal one (see Fig. 5.16). All complete data synchronizers, standard and custom, are detected and verified with the enabler-based synchronizer model. This removes the need for identifying and verifying them through handshake or FIFO models. Partial data synchronizers are no longer verified due to the fact that the structure is incomplete.

Figure 5.16 presents the CDC verification flow integrating the enabler-based synchronizer model. Compared to the current verification flow of Fig. 5.11, the user-provided constraints have disappeared: they are no longer needed, neither for the structural identification nor for the functional verification. With our proposed model, we achieve a fully automatic data synchronizer verification flow.

5.5   Practical Experiments

The RTL design used for practical experiments includes a data synchronizer that falls into the generic model of Fig. 5.14, where the FIFO size is a parameter that influences the synchronization model complexity. The synchronizer is interfaced with two handshakes, in the source and destination clock domains respectively. Those external components are used to control the data transfer from the source clock domain to the synchronizer and from the synchronizer to the destination clock domain.

Figure 5.17 shows the block diagram of the RTL design used for practical experiments. It includes a custom synchronizer, two handshake structures at the interface noted HS, and the counter noted CNT in the fan-in of the source clock domain data path.

The practical experiments that relate to the structural detection of data synchronizers have been made with three industrial CDC verification tools: one static CDC verification tool (say Tool 1), a CDC checker tool that embeds a model checking engine allowing properties specification by the user (say Tool 2) and a CDC tool dedicated to structural verification (say Tool 3). The objective is to exhibit, for each tool, the type of structure detected as a function of the data synchronizer types allowed. The results are reported in Table 5.1.

The first three columns give the values of the (tool) parameters enabling the detection of the three following data synchronizers: handshake, FIFO and control-based. For instance, in the first line, only control-based synchronizers are looked for in the RTL architecture, whereas in line 2 the recognition of FIFO's and control-based synchronizers is set. The last line forces the tools to detect handshakes only.

Columns 4, 5 and 6 give the results obtained with Tool 1, Tool 2 and Tool 3, showing the type of synchronizer found by the tool after pattern matching.

The pattern matching algorithms implemented in the three CDC tools return quite distinct results.

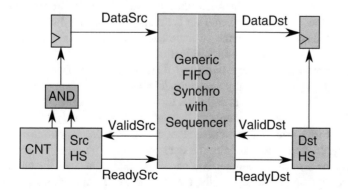

Fig. 5.17 Test case architecture

Table 5.1 Structural detection results

Handshake	FIFO	Control-based	Tool 1	Tool 2	Tool 3
0	0	1	Control-based	Control-based	Control-based
0	1	1	Half-FIFO	Control-based	Half-FIFO
1	0	1	Control-based	Handshake	Handshake
1	1	1	Half-FIFO	Handshake	Half-FIFO
0	1	0	Half-FIFO	Unsynchronized	Half-FIFO
1	1	0	Half-FIFO	Handshake	Half-FIFO
1	0	0	Unsynchronized	Handshake	Handshake

Tool 1 gives higher priority to FIFO detection. Since our structure is not a standard FIFO, the synchronizer is not well detected; this is noted "Half-FIFO" in the table. Also, Tool 1 is not able to detect a handshake structure even if all other types of data synchronizer are disabled. So the best solution to perform a complete CDC verification flow with Tool 1 is to only allow partial detection as control-based synchronizer.

Tool 2 gives higher priority to handshake detection with interesting results, so the synchronizer is detected like an enabler-based synchronizer. Tool 2 is not able to detect a FIFO structure if all other types of data synchronizer are disabled, which is a normal behavior since our structure is not a standard FIFO. So Tool 2 allows to perform a complete CDC verification flow when all synchronizer structures are set.

Tool 3 gives higher priority to FIFO detection. As for Tool 1, since our structure is not a standard FIFO, the synchronizer is not well detected (Half-FIFO). Tool 3 is able to detect a handshake structure if FIFO identification is disabled. So, the best solution to perform a complete CDC verification flow is to disable FIFO detection.

The results show that, using three independent industrial CDC checker tools, the synchronizer model proposed has been identified (through a handshake scheme) by two of the three tools, provided we set the execution with an appropriate configuration.

As a side result, we benchmarked the three industrial CDC checker tools, and showed that the second one is better for the detection of data synchronizers.

In our methodology, after the structural identification of the synchronizers in the flat RTL design, the second verification phase is model checking. We have performed practical experiments on the RTL design with again three industrial verification tools: two general purpose model checkers (say Tool 4 and Tool 5) and a specialized CDC checker that embeds a model checking engine, our previous Tool 2. The design includes synchronizers that fall into the generic model of Fig. 5.14, where the FIFO size is a parameter that influences the synchronization model complexity. In the source domain, the data path in the fan-in of DataSrc contains a counter; the size of the counter is a parameter that influences the complexity of the source domain data path. Finally, the bit width of DataSrc is also generic. The source clock cycle is $2/3$ of the destination clock cycle.

The results are reported in Table 5.2. The first three columns give the values of the three generic parameters: FIFO depth, counter size (in bits), and DataSrc width.

Columns 4, 6 and 8 give the results obtained with Tool 4, Tool 2 and Tool 5 on the properties associated to our synchronization protocol (see Sect. 5.4.3), i.e. four properties per FIFO entry, with the proof time in seconds. Columns 5, 7 and 9 give the results obtained with Tool 4, Tool 2 and Tool 5 on the properties associated to the partial recognition of the FIFO (see Sect. 5.3.2), i.e. there is only one property per FIFO entry, with the proof time in seconds. In these last six columns, T.O. stands for "time out", which means that the tool has neither found a proof nor a counter example in the maximal time set for the experiment: 300 s.

The table is divided in three parts, that correspond to the variation of one among the three parameters.

In the first experiment, the FIFO depth takes all even values between 4 and 20. The complexity of the WrFSM and RdFSM is directly impacted by this parameter.

If we compare the stability results, we see that Tool 4 stops getting a proof on a FIFO depth of 10, while Tool 2 stops at 16 and Tool 5 stops at 20. Our model does not bring any improvement, because the two FSMs are part of the synchronizer model.

The second experiment has a fixed FIFO size of 4 and varies the counter size from 1 bit to the size when a time-out appears. In the case of partial recognition, Tool 4 stops getting a proof on 7 bits; Tool 2 provides better performances, and a proof is obtained for up to 119 bits; Tool 5 provides the best performances since a proof is obtained for up to 10,000 bits. With our model, the counter is not taken in the fan-in cone of DataSrc, and thus a proof is obtained for all sizes and with the same effort (see Columns 4, 6, and 8).

The last experiment shows that all synchronization models are independent of the data width for Tools 4 and 2. However, with Tool 5, in case of partial recognition, the proof runtime increments with the data width but with very good performance, since a proof is obtained for up to 2500 bits. So, we can assume that this tool is able to limit the proof runtime for data stability analyzing both the source logic outside the synchronizer (counter size) and the crossing data path (data width).

The results show that, using three independent industrial formal engines, the properties associated to our model are easier to verify than those related to partial synchronizers. As a side result, we benchmarked three industrial verification tools, and showed that the third one is better for the verification of properties related to CDC although it is not a specialized CDC checker tool.

If we come back to the CDC verification flow, we can see that, using Tool 2, we can allow the detection of our custom data synchronizer as a handshake. However, concerning the functional verification, the full data synchronizer detection has no influence on the proof runtime. This means that the properties verified by Tool 2 deal directly with the data (see properties in Sect. 5.3.2) and not with the Enable signal (properties in Sect. 5.4.3), despite the fact that this signal has been recognized correctly in the structural identification phase. A solution can be to generate the properties associated to the enabler-based synchronizer model from the handshake scheme detected by the tool.

Table 5.2 Formal verification results

FIFO depth	Counter size	Data width	Protocol results 4	Stability results 4	Protocol results 2	Stability results 2	Protocol results 5	Stability results 5
4	2	50	Pass (11)	Pass (14)	Pass (1)	Pass (2)	Pass (0.3)	Pass (1.6)
6	2	50	Pass (48)	Pass (32)	Pass (1)	Pass (2)	Pass (0.7)	Pass (2.1)
8	2	50	Pass (192)	Pass (140)	Pass (2)	Pass (3)	Pass (0.8)	Pass (2.8)
10	2	50	T.O. (300)	T.O. (300)	Pass (4)	Pass (4)	Pass (1.4)	Pass (4.7)
12	2	50	T.O. (300)	T.O. (300)	Pass (5)	Pass (7)	Pass (2.1)	Pass (11.5)
14	2	50	T.O. (300)	T.O. (300)	Pass (290)	Pass (295)	Pass (3.9)	T.O. (300)
16	2	50	T.O. (300)	T.O. (300)	T.O. (300)	T.O. (300)	Pass (7.8)	T.O. (300)
18	2	50	T.O. (300)	T.O. (300)	T.O. (300)	T.O. (300)	Pass (15)	T.O. (300)
20	2	50	T.O. (300)	T.O. (300)	T.O. (300)	T.O. (300)	Pass (40)	T.O. (300)
22	2	50	T.O. (300)	T.O. (300)	T.O. (300)	T.O. (300)	Pass (60)	T.O. (300)
24	2	50	T.O. (300)	T.O. (300)	T.O. (300)	T.O. (300)	Pass (205)	T.O. (300)
26	2	50	T.O. (300)	T.O. (300)	T.O. (300)	T.O. (300)	T.O. (300)	T.O. (300)
4	1	50	Pass (11)	Pass (7)	Pass (1)	Pass (1)	Pass (0.3)	Pass (0.9)
4	2	50	Pass (11)	Pass (14)	Pass (1)	Pass (2)	Pass (0.3)	Pass (1.7)
4	3	50	Pass (11)	Pass (23)	Pass (1)	Pass (2)	Pass (0.3)	Pass (1.9)
4	4	50	Pass (11)	Pass (36)	Pass (1)	Pass (2)	Pass (0.3)	Pass (2.1)
4	5	50	Pass (11)	Pass (154)	Pass (1)	Pass (3)	Pass (0.3)	Pass (2.7)
4	6	50	Pass (11)	Pass (259)	Pass (1)	Pass (3)	Pass (0.3)	Pass (4.2)
4	7	50	Pass (11)	T.O. (300)	Pass (1)	Pass (3)	Pass (0.3)	Pass (5.6)
4	8	50	Pass (11)	T.O. (300)	Pass (1)	Pass (3)	Pass (0.3)	Pass (8.4)
4	9	50	Pass (11)	T.O. (300)	Pass (1)	Pass (4)	Pass (0.3)	Pass (31)
4	10	50	Pass (11)	T.O. (300)	Pass (1)	Pass (4)	Pass (0.3)	Pass (31)
4	32	50	Pass (11)	T.O. (300)	Pass (1)	Pass (7)	Pass (0.3)	Pass (31)
4	64	50	Pass (11)	T.O. (300)	Pass (1)	Pass (10)	Pass (0.3)	Pass (31)
4	118	50	Pass (11)	T.O. (300)	Pass (1)	Pass (20)	Pass (0.3)	Pass (35)
4	119	50	Pass (11)	T.O. (300)	Pass (1)	T.O. (300)	Pass (0.3)	Pass (35)
4	500	50	Pass (11)	T.O. (300)	Pass (1)	T.O. (300)	Pass (0.3)	Pass (58)
4	1000	50	Pass (11)	T.O. (300)	Pass (1)	T.O. (300)	Pass (0.3)	Pass (75)
4	2000	50	Pass (11)	T.O. (300)	Pass (1)	T.O. (300)	Pass (0.3)	Pass (115)
4	5000	50	Pass (11)	T.O. (300)	Pass (1)	T.O. (300)	Pass (0.3)	Pass (123)
4	10, 000	50	Pass (11)	T.O. (300)	Pass (1)	T.O. (300)	Pass (0.3)	Pass (190)
4	2	100	Pass (11)	Pass (14)	Pass (1)	Pass (2)	Pass (0.3)	Pass (3.6)
4	2	200	Pass (11)	Pass (14)	Pass (1)	Pass (2)	Pass (0.3)	Pass (16)
4	2	300	Pass (11)	Pass (14)	Pass (1)	Pass (2)	Pass (0.3)	Pass (21)
4	2	500	Pass (11)	Pass (14)	Pass (1)	Pass (2)	Pass (0.3)	Pass (30)
4	2	1000	Pass (11)	Pass (14)	Pass (1)	Pass (2)	Pass (0.3)	Pass (58)
4	2	2500	Pass (11)	Pass (14)	Pass (1)	Pass (2)	Pass (0.3)	Pass (184)
4	2	5000	Pass (11)	Pass (14)	Pass (1)	Pass (2)	Pass (0.3)	T.O. (300)

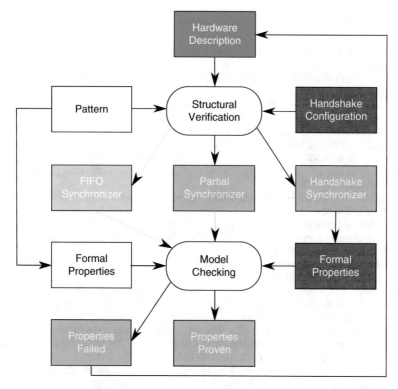

Fig. 5.18 Experimented data synchronizers verification flow

Figure 5.18 presents the CDC verification flow integrating the enabler-based synchronizer model practically put in place. It shows that the user-based constraints are no longer needed, neither for the structural identification nor for the functional verification. So with the proposed model we come back to a quasi-automatic data synchronizer verification flow: only the formal functional properties to be model checked must be provided by the user.

5.6   Conclusion

We have shown that industrial CDC verification tools fail to be efficient on a flat design integrating custom synchronizers: these are only partially recognized with a standard configuration and the properties that are model checked are not precise enough, leading to inconclusive results. We proposed a generic synchronization model enabling the extraction of a greater variety of synchronizers, and its associated set of properties which limit the design area to be model checked. The adoption of our model allows for faster verification turn around time, and reduces human intervention, which is critical in a competitive industrial environment.

References

1. Burns, F., Sokolov, D., Yakovlev, A.: GALS synthesis and verification for xMAS models. In: DATE (2015)
2. Cadence: Conformal CDC (2015). URL http://www.cadence.com/products/ld/constraint_designer/pages/default.aspx
3. Chaney, T.: My Work on All Things Metastable OR: (Me and My Glitch). Blendics White Paper (2012)
4. Chaney, T., Molnar, C.: Anomalous behavior of synchronizer and arbiter circuits. IEEE Trans. Comput. **C-22**(4), 421–422 (1973). DOI 10.1109/T-C.1973.223730
5. Chaturvedi, S.: Static analysis of asynchronous clock domain crossings. In: Design, Automation Test in Europe Conference Exhibition (DATE), 2012, pp. 1122–1125 (2012). DOI 10.1109/DATE.2012.6176664
6. Foster, H., Group, W.: IEEE standard for property specification language (PSL). pub-IEEE-STD-1850 (2010)
7. Ginosar, R.: Fourteen ways to fool your synchronizer. In: Proceedings of the Ninth International Symposium on Asynchronous Circuits and Systems, 2003, pp. 89–96 (2003). DOI 10.1109/ASYNC.2003.1199169
8. Ginosar, R.: Metastability and synchronizers: a tutorial. IEEE Des. Test Comput. **28**(5), 23–35 (2011). DOI 10.1109/MDT.2011.113
9. Kapschitz, T., Ginosar, R.: Formal verification of synchronizers. In: Borrione, D., Paul, W. (eds.) Correct Hardware Design and Verification Methods, LNCS, vol. 3725, pp. 359–362. Springer, Berlin (2005). DOI 10.1007/11560548_31
10. Karimi, N., Chakrabarty, K.: Detection, Diagnosis, and Recovery From Clock-Domain Crossing Failures in Multiclock SoCs. IEEE Trans. Comput. Aided Des. Integr. Circuits Syst. **32**(9), 1395–1408 (2013). DOI 10.1109/TCAD.2013.2255127
11. Leong, C., Machado, P., Bexiga, V., Teixeira, J.P.: Built-in clock domain crossing (CDC) test and diagnosis in GALS systems. In: Proceedings of DDECS 2010, pp. 72–77 (2010). DOI 10.1109/DDECS.2010.5491815
12. Li, B., Kwok, C.K.: Automatic formal verification of clock domain crossing signals. In: Design Automation Conference, 2009. ASP-DAC 2009. Asia and South Pacific, pp. 654–659 (2009). DOI 10.1109/ASPDAC.2009.4796554
13. Lodhi, F., Hasan, S., Sharif, N., Ramzan, N., Hasan, O.: Timing variation aware dynamic digital phase detector for low-latency clock domain crossing. IET Circuits Dev. Syst. **8**(1), 58–64 (2014). DOI 10.1049/iet-cds.2013.0067
14. Mentor Graphics: Questa CDC (2015). URL http://www.mentor.com/products/fv/questa-cdc
15. Schmaltz, J.: A formal model of clock domain crossing and automated verification of time-triggered hardware. In: Formal Methods in Computer Aided Design, 2007. FMCAD '07, pp. 223–230 (2007). DOI 10.1109/FAMCAD.2007.22
16. Synopsys: SpyGlass CDC (2015). URL http://www.synopsys.com/Tools/Verification/static-formal-verification/Pages/spyglass-cdc.aspx
17. Verma, S., Dabare A.S.: Design how-to understanding clock domain crossing issues. EETimes (2007). URL http://www.eetimes.com/document.asp?doc_id=1276114

Chapter 6
Temporal Decoupling with Error-Bounded Predictive Quantum Control

Georg Gläser, Gregor Nitsche, and Eckhard Hennig

6.1 Introduction

Virtual prototyping and mixed-signal mixed-level simulation are commonly used in the design of smart-sensor systems for early firmware development and system-level hardware/software co-verification. A co-simulation environment for analog and digital components—including an analog simulator, a digital simulator and a microcontroller instruction-set simulator (ISS)—provides a base for the concurrent development of circuit hardware and signal processing algorithms (Fig. 6.1). In the following, we consider the case of a smart-sensor system where a timing-accurate verification of signal processing algorithms and data communication protocols requires full-system simulations over a large number of signal acquisition cycles. Each acquisition cycle involves many interactions of the microcontroller with the analog part, such as configuring a programmable gain amplifier (PGA), triggering an analog-to-digital converter (ADC), waiting for the A-to-D-conversion to complete and receiving the converted data (Fig. 6.2).

To simulate such a mixed-signal design in an early stage of development efficiently, including its firmware, we use the *temporal decoupling* (TD) approach

G. Gläser (✉)
IMMS Institut für Mikroelektronik- und Mechatronik-Systeme gemeinnützige GmbH,
Ehrenbergstr. 27, 98693 Ilmenau, Germany
e-mail: georg.glaeser@imms.de

G. Nitsche
OFFIS - Institut für Informatik Oldenburg, Escherweg 2, 26121 Oldenburg, Germany
e-mail: gregor.nitsche@offis.de

E. Hennig
Reutlingen University, Alteburgstr. 150, 72762 Reutlingen, Germany
e-mail: eckhard.hennig@reutlingen-university.de

© Springer International Publishing Switzerland 2016
R. Drechsler, R. Wille (eds.), *Languages, Design Methods, and Tools
for Electronic System Design*, Lecture Notes in Electrical Engineering 385,
DOI 10.1007/978-3-319-31723-6_6

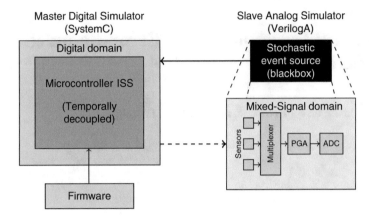

Fig. 6.1 Virtual prototype structure of a smart-sensor system: a microcontroller ISS interacting with an analog frontend simulated in a coupled simulator. For the aspect of the event synchronization the analog frontend can be considered as a stochastic black box event source

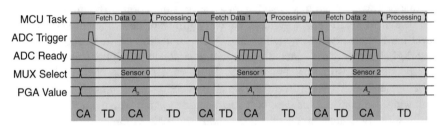

Fig. 6.2 Timing diagram for a full conversion cycle of a three sensor multi-sensor system as it is seen in the digital simulator

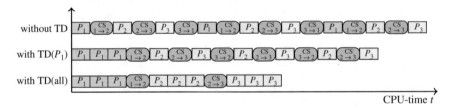

Fig. 6.3 Execution of three processes without TD, with TD of a single process (P_1) and with TD of all processes: the reduced number of context switches in TD simulations leads to a reduced execution time in the simulator CPU

described in the SystemC LRM [1] and its extension to a *cycle-count accurate* simulation (CCA) that preserves clock-edge timing accuracy at time-quantum boundaries [2]. A speed gain over cycle-based simulation can thus be achieved by reducing the number of inter-process synchronization points in the simulation time interval at the expense of timing accuracy inside the time quantum (Fig. 6.3).

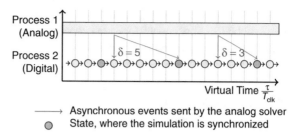

Fig. 6.4 Event processing latency δ in mixed signal simulations caused by temporal decoupling

As a result, an asynchronous external event sent to the SystemC core from a coupled analog simulator is processed, in general, with some extra delay δ because it can only be handled at the next synchronization point (Fig. 6.4). The resulting timing error may be of serious consequence for the functional correctness of the simulation result: consider a microcontroller that checks for incoming requests from an ADC only within fixed time slots. Excessive latency in event handling might cause the simulation model of the digital backend to miss important input and, therefore, feed the subsequent data processing chain with incomplete or wrong data.

Returning to fully cycle-accurate lock-step simulation (CA) would solve this problem; however, this is undesirable for performance reasons. At the same time, it is not even necessary because cycle-accurate simulation is only required for time intervals where the simulator interacts with another process. Figure 6.2 shows that there are still extended time intervals TD in between these phases of interaction where temporal decoupling of mixed-signal and firmware simulation processes is possible without loss of system-level timing accuracy. It follows that an optimal compromise between performance and accuracy can be achieved if CA and TD simulation methods are combined in the simulation environment and applied selectively to time intervals with and without process interaction. This requires a method that continuously tracks and predicts the permissible TD windows while the simulation is running.

In this paper, we present a *predictive* temporal decoupling approach that minimizes external event processing latency so that remaining latencies are either within acceptable limits or can be repaired with reasonable overall cost by checkpointing and rollback methods [3]. To estimate the position of future events in time and to synchronize our simulation at the correct points in time, we use a linear prediction scheme. In addition, we introduce a safety margin to adapt our method to the stochastic properties, i.e. the *randomness*, of the event source, thus making the approach applicable to more general use cases.

In Sect. 6.2, we summarize the related work on temporal decoupling for faster hardware/software co-verification. Section 6.3 gives an overview of the fundamental concepts relevant to our work. The details of our *Predictive Temporal Decoupling* (PTD) approach are explained in Sect. 6.4. In Sect. 6.5, we analyze the theoretical performance of PTD by deriving a set of cost functions that express the expected simulation effort in terms of key parameters such as time quantum size and CPU time per simulation cycle. Section 6.6 shows the results of simulating an abstraction of the real smart-sensor system, using the optimal time quanta (TQ) calculated by our method. Finally, conclusions are presented in Sect. 6.7, including a perspective for future work on the subject of this paper.

6.2   Related Work

Delayed processing of asynchronous events is a problem which has been discussed before by Damm et al. in the context of SystemC-AMS modules linked to temporally decoupled digital models that are simulated with fixed time quanta [4]. If incoming events are generated deterministically from within the SystemC simulation, it is possible to calculate the position of the next event in time and add a synchronization point at this position to obtain a timing-accurate simulation result [2, 5, 6].

Annotation of known timings to ensure proper synchronization has been suggested by Salimi Khaligh and Radetzki [7]. Schirner and Domer propose the introduction of further delays to correct the timing of bus transfers [8].

An overview of recently published methods in the context of real-time embedded software simulation is given by Bringmann et al. [9]. This paper also discusses the consequences of delayed processing of asynchronous interrupts. However, none of the above approaches take asynchronous events from outside of the simulator into account where the corresponding points in time are not known in advance.

Other proposed methods try to correct the latency by going back in simulated time using checkpointing and rollback mechanisms [3, 10]. Rollback methods are very useful for solving the TD synchronization problem, but they require deep changes in the simulation engine and may not improve overall simulation performance if applied without tuning to the specific timing characteristics of the SystemC model to be simulated.

6.3   Temporal Decoupling

6.3.1   Fundamentals

In a digital simulator, a scheduler synchronizes the simulated processes after each clock cycle or after each event, depending on the time granularity of the underlying simulation method [1]. Every synchronization requires the host processor to perform *context switching* (CS) between processes, which entails switching effort in the scheduler [2, 5]. In lock-step synchronization mode, context switching consumes a substantial share of the available CPU resources.

Therefore, simulation performance would benefit considerably from reducing the number of CS phases, which can be achieved by *temporal decoupling* (TD).

The idea behind temporal decoupling, shown in Fig. 6.3, is to simulate one or more processes independently of the other processes for an extended period of time that spans an arbitrary multiple of the fundamental clock period. This period is called a *time quantum* (TQ). If there is no inter-process communication between decoupled processes within the TQ, synchronization after every process time step is not required and the corresponding context switching effort can be saved to improve simulation speed.

The most significant performance gain is achieved by temporally decoupling all processes. As long as the above assumption regarding inter-process communication is satisfied, or if synchronization delays introduced by TD have no influence on system behavior, the simulation result remains correct.

The predictive temporal decoupling method presented in the following sections is based on this general principle. However, in contrast to the standard TD approach, we define our TQ in terms of integer multiples of clock cycles rather than absolute time periods. Thus, time quanta are aligned with the system clock, which helps to preserve clock-edge timing accuracy at time quantum boundaries [2].

6.3.2   Problem Statement and Generalized Model

The synchronization scheme in a TD simulation assumes that no inter-process dependencies exist within a TQ. If this assumption is violated by an event that should be processed immediately, a possible latency δ is introduced because the event can only be processed after synchronizing the simulation in the following TQ (Fig. 6.5). Let Process 1 represent an external blackbox entity coupled to the SystemC simulator, such as an analog solver, a compiled IP module or a hardware-in-the-loop component. The events generated from Process 1 introduce an inter-process dependency which can only be dealt with in a future timestep. Since Process 2 is executed with TD, it assumes dependence on only its internal state and the inputs from the beginning of the current TQ. Therefore, arbitrary inter-process

Fig. 6.5 Event processing latency δ in temporally decoupled simulations caused by fewer synchronizations due to the size of the time quantum

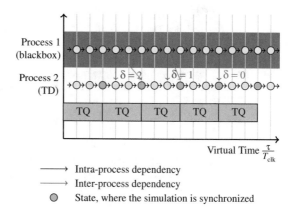

communication events cannot be handled exactly at their time of occurrence, which introduces a processing latency up to the beginning of the next TQ. As a more abstract model, in this paper, every process not inspectable from the SystemC simulator is considered as a blackbox which generates events with stochastic timing properties.

Event processing latency may affect both timing and functional correctness of a simulation. If further knowledge about the simulated model is available, errors can be avoided by inserting additional synchronization points [2, 9]. Since this is not possible for events generated outside of the SystemC simulation context, these events are still not processed correctly. In our application scenario, the mixed-signal system contains event sources whose properties are not visible from within the SystemC simulation context. Therefore, a normal TD simulation with uncorrelated TQ would likely produce invalid simulation results.

An obvious approach to solve this problem would be to shrink the size of the TQ to ensure that the distance to the next synchronization point is sufficiently small (Fig. 6.6). However, reducing TQ size adversely affects simulation speed [4, 5]. Moreover, a small TQ alone does not guarantee elimination of latencies. In addition to choosing an optimal TQ size, the TQ boundaries have to be aligned with the incoming events.

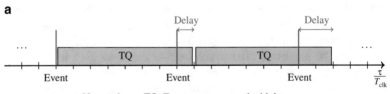

Non-optimum TQ: Events are processed with latency.

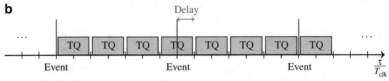

Non-optimum TQ: More synchronizations than necessary slow down the simulation.

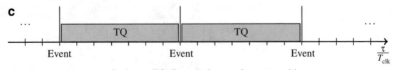

Optimum TQ: Boundaries match event positions.

Fig. 6.6 The TQ should be selected such that the boundaries meet the positions of events to ensure correct handling by the receiving process: in case (**a**), the event distances and positions do not match the TQ boundaries. Reducing the TQ size in case (**b**) is not sufficient since the position has also to be synchronized as shown the last case (**c**)

6.4   Predictive Temporal Decoupling

To calculate a TQ that minimizes possible latencies, we use the generic model of two communicating processes shown in Fig. 6.5. In many relevant cases, such as desynchronization phenomena or quasi-periodic transmission of events from a sensor ADC, the underlying stochastic processes can be assumed to be stationary. The position of the next event can thus be predicted with adaptive signal processing methods [11]. In the following, we model the blackbox process as a stochastic source which generates a random sequence of events with gaussian distribution of the time intervals.

The optimal size and placement of the next TQ can be calculated with an adaptive filter algorithm. As shown in Fig. 6.7 [12], we use a time-discrete *Wiener-filter* predictor as an example for proving the concept of predictive temporal decoupling (PTD). The values to be computed by the filter are the estimated intervals $\hat{T}_i$ between two consecutive events from the blackbox source.

In the following, the filter algorithm is described only briefly; it is discussed in more detail in [11] and [12]. The algorithm tries to minimize the prediction error by optimizing the weights of an FIR filter (Fig. 6.8). It assumes that the model order, i.e. the length of the FIR filter, and the mean value μ of the input data sequence T_i is known. Hence, we have to estimate the mean μ and the autocorrelation matrix $\mathbf{R}$ between consecutive event distances T_i and $T_{(i+1)}$.

Since the implemented Wiener filter expects the sequence of event intervals to have zero mean, we first subtract the estimated mean from all samples. The results are combined in a vector $\mathbf{u}$, which is used in the following matrix calculations.

$$\mathbf{u} = [(T_N - \mu)\ (T_{N-1} - \mu) \ldots (T_1 - \mu)]^{\mathrm{T}} \tag{6.1}$$

Fig. 6.7 General structure of a simple adaptive filter: an algorithm adapts the weights of an FIR filter $\mathbf{w}$ to minimize the prediction error e_n

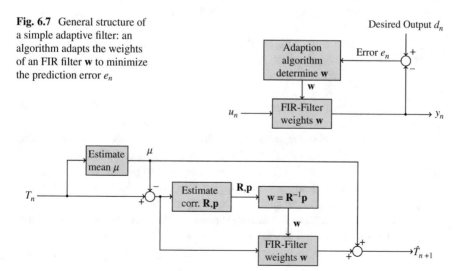

Fig. 6.8 Structure of the predictor used to estimate the time until the next event will most likely arrive $\hat{T}_{n+1}$

The weight vector **w** of the predictor is calculated using the following cost function, which represents the error at the filter output.

$$J(\mathbf{w}) = \mathbb{E}\left\{|y_N - d_N|^2\right\} = \mathbb{E}\left\{|\mathbf{w}^H\mathbf{u} - T_{N+1}|^2\right\} \tag{6.2}$$

Where y_N denotes the output value computed by the filter and d_N the desired value $T_{(N+1)}$. This function is minimized with respect to **w**, resulting in the *Wiener-Hopf* equation [12]

$$\mathbf{w} = \mathbf{R}^{-1}\mathbf{p} \tag{6.3}$$

In Eq. (6.3), **R** denotes the auto-correlation matrix of **u** and **p** the cross-correlation between the input data and the expected output data:

$$\mathbf{R} = \mathbb{E}\left\{\mathbf{uu}^H\right\} \tag{6.4}$$

$$\mathbf{p} = \mathbb{E}\left\{\mathbf{u} \cdot T_{N+1}\right\} \tag{6.5}$$

To estimate **R** and **p**, we use a standard estimator with a forgetting factor [11]. If the model order is known, the matrix **R** is invertible. With the resulting weight vector **w**, the FIR filter can predict the length of the next event interval:

$$\hat{T} = \mathbf{u}^T\mathbf{w} \tag{6.6}$$

During a simulation, the filter parameters are continually updated with the timing information from each received event. If the process is stationary, the estimates for **R** and **p** converge, resulting in constant filter weights **w**.

Due to the stochastic properties of the blackbox process, some uncertainty still remains in the predicted value $\hat{T}$. In fact, the next event may occur just a bit before the predicted point in time. This results in residual processing latency, which may be acceptable in a given situation. In the worst case, however, the event occurs just after a new time quantum has been started, resulting in the longest possible delay for the current TQ size.

To account for stochastic uncertainties, we combine TD simulation with cycle-accurate simulation: assume that the system has been simulated in TD mode up to the predicted time of the next event, but the event has not occurred yet. In this case, we proceed with cycle-accurate simulation until the expected event has arrived. Subsequently we resume with TD simulation using the next predicted TQ. As shown in Fig. 6.9, this allows all late events to be handled accurately. If the event timing distribution is symmetric about zero, this amounts to 50% of all events.

Assuming a gaussian distribution with standard deviation σ, we can deal with the majority of early events in a similar fashion: by synchronizing our system already at $\left(\hat{T} - S \cdot \sigma\right)$ and simulating cycle-accurately from this earlier point in time, we

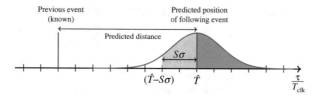

Fig. 6.9 The simulation is done cycle-accurately starting from the point $(\hat{T} - S\sigma)$ that includes the predicted position $\hat{T}$ and a safety margin $S\sigma$

can guarantee exact handling for all events with a timing variation within the safety margin $S\sigma$. The probability p_{miss} of missing an event can be calculated with the error function $Q(x)$ [12].

$$p_{\text{miss}} = Q(S) \tag{6.7}$$

The safety margin parameter S can be chosen so as to yield an acceptable miss rate.

Real-world systems frequently exhibit quasi-periodic behaviour, which results in bounded error distributions f_{error}.

$$f_{\text{error}}(\tau) = \begin{cases} f(\tau) \neq 0 \text{ for } |\tau| < \tau_0 \\ f(\tau) = 0 \text{ for } |\tau| \geq \tau_0 \end{cases} \tag{6.8}$$

In this case, the safety margin can be determined such that no events are missed by our simulation approach. Hence, it is possible to eliminate timing errors completely for this practically relevant scenario.

As a special case of a bounded distribution, periodic events can be modeled as a zero-variance stochastic process with a mean μ that is equal to the period. Therefore, after a certain time for estimating the mean and correlation properties of the event sequence, an appropriate TQ can be calculated such that every event is handled at its exact position in time and the simulation is always cycle accurate.

In the worst case of completely random and, therefore, unpredictable event intervals, the safety margin $S\sigma$ is maximized to the extreme case where only cycle-accurate simulation is performed. Thus, incoming events are still processed correctly but at the expense of overall simulation performance.

6.5 A Cost Model for Performance Evaluation of SystemC Simulation Strategies

Evaluating different model simulation and synchronization strategies against each other requires a metric for the corresponding simulation performance. Until now, Nitsche et al. [2] and Damm et al. [4] measured the performance of temporally decoupled simulations by observing the effects of varying the TQ. However, the literature does not provide a unified way of comparing existing techniques.

In this section, we derive a set of cost functions that model the expected CPU effort for a given simulation approach and parameter configuration. The presented functions are based on the cost estimation model for Discrete Event System (DES) simulation with optimistic scheduling and rollback proposed by Lin et al. [13].

Our cost model contains parameters that may not be known a priori and depend strongly on the implementation and the instantaneous state of the model. To eliminate the influence of the model state, all parameters denote *average* simulation costs over a complete simulation period. Furthermore, the parameters are deeply influenced by the compiler and the executing processor, since they are defined with respect to CPU time. Hence, it is necessary to provide information about the simulation environment in which the parameters were measured to permit proper comparison of experimental results.

6.5.1 Cycle-Accurate Simulation

Cycle-accurate simulation ensures timing accuracy at clock-cycle boundaries. To keep causality, the system must be synchronized after each simulated cycle. The simulation of a single process is depicted in Fig. 6.10, which introduces the cost factors K_{sim} and K_{sync} graphically. The required number of synchronizations N is given by the number of simulated cycles. The average CPU effort needed to simulate a single cycle in a model—without synchronization—is denoted by K_{sim}. Likewise, K_{sync} denotes the average CPU time needed for one synchronization of the model. The total cost of simulating N cycles can thus be expressed as follows.

$$K_{total}^{CA}(N) = N \cdot K_{sim} + N \cdot K_{sync} \qquad (6.9)$$

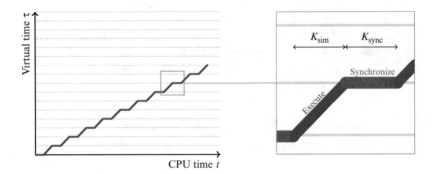

Fig. 6.10 Cycle-accurate simulation of a model

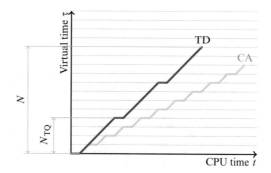

Fig. 6.11 Temporal decoupled simulation of a model. The model is not synchronized after every single cycle

6.5.2 Temporally Decoupled Cycle-Accurate Simulation

Contrary to cycle-accurate simulation, a temporally decoupled model is synchronized only after the simulation of a complete time quantum (Fig. 6.11). Using the same notation as in Sect. 6.5.1, the simulation cost for a single TQ of N_{TQ} cycles is defined as

$$K_{TQ}^{TD} = N_{TQ} \cdot K_{sim} + K_{sync} \tag{6.10}$$

To calculate the total simulation cost for N cycles, it is assumed—without loss of generality—that N is an integer multiple of the length N_{TQ} of the TQ.

$$K_{total}^{TD}(N) = \sum_{i=0}^{n} K_{TQ}^{TD} = N \cdot K_{sim} + \frac{N}{N_{TQ}} \cdot K_{sync} \tag{6.11}$$

As can be expected, for $N_{TQ} = 1$, the cost model for TD simulation is consistent with the cost model for cycle-accurate simulation.

6.5.3 Temporally Decoupled Simulation with Rollback

A method to ensure causality and accuracy in temporally decoupled simulation is a *rollback* (RB) mechanism as shown in Fig. 6.12. Rollback allows a simulation to return to a previously saved state and resimulate from there to correct errors resulting from some previously missed event. With this optimistic synchronization method, it is possible to eliminate event processing latencies. However, simulators that support this feature are not widely available yet. Still, the concept is evaluated here for comparison with the proposed predictive temporal decoupling approach. First steps towards implementing a rollback mechanism have been done by Monton et al. [3].

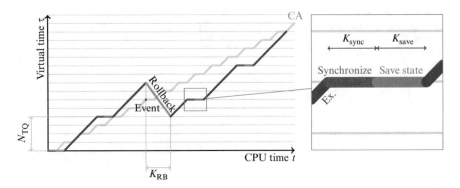

Fig. 6.12 Temporal decoupled simulation with rollback. The simulation is rolled back to react to a missed event

The derivation of a cost function for this simulation strategy is more complex than the previous ones. We assume that every synchronization is followed by saving the current simulation state in order to be prepared for a possible rollback after the next TQ. At the end of each TQ, the simulator checks whether a rollback is required to capture a missed event. In a first step, consider a TQ where no event has been missed. In this case, only the effort needed to save a state K_{save} is added to the simulation cost for the simulation of a TQ.

$$K_{\text{TQ},1}^{\text{TD,RB}} = N_{\text{TQ}} \cdot K_{\text{sim}} + K_{\text{sync}} + K_{\text{save}} \qquad (6.12)$$

Secondly, consider a TQ during which an external event is received. After the end of the TQ has been reached, the simulation results are discarded and the model state is rolled back to the beginning of the TQ, resulting in additional cost K_{RB}. Afterwards, the time interval is resimulated taking the missed event into account. Then, the model is synchronized at this point and a new state is saved. The cost for this sequence is

$$K_{\text{TQ},2}^{\text{TD,RB}} = 2N_{\text{TQ}} \cdot K_{\text{sim}} + K_{\text{RB}} + K_{\text{sync}} + K_{\text{save}} \qquad (6.13)$$

To determine the expected simulation cost for a TQ, it is necessary to define p_{RB} as the probability that a rollback is necessary. This quantity depends obviously on the size of the TQ, since the probability that an event arrives within a time interval grows with increasing size of the interval. Therefore, the expected cost for a single TQ is given as

$$K_{\text{TQ}}^{\text{TD,RB}} = (1 - p_{\text{RB}}) \cdot K_{\text{TQ},1}^{\text{TD,RB}} + p_{\text{RB}} \cdot K_{\text{TQ},2}^{\text{TD,RB}} \qquad (6.14)$$

Simplifying this expression yields

$$K_{\text{TQ}}^{\text{TD,RB}} = (1 + p_{\text{RB}}) \cdot N_{\text{TQ}} \cdot K_{\text{sim}} + p_{\text{RB}} \cdot K_{\text{RB}} + K_{\text{sync}} + K_{\text{save}} \qquad (6.15)$$

In a next step, the total simulation cost is calculated assuming again that the number of simulated cycles N is an integer multiple of the size of the time quantum N_{TQ}.

$$K_{total}^{TD,RB}(N) = (1 + p_{RB}) \cdot N \cdot K_{sim} + \frac{N}{N_{TQ}} (p_{RB} \cdot K_{RB} + K_{sync} + K_{save}) \qquad (6.16)$$

These expressions are similar to the ones obtained in the previous section for TD simulations without rollback. If no rollbacks are necessary ($p_{RB} = 0$) and if the additional saving effort is not significant, the expressions become identical.

To obtain an expression that depends on the probability p_{event} that an event occurs in a certain clock cycle—assuming uniform distribution over the TQ—the following expression for p_{RB} is substituted in Eq. (6.16).

$$p_{RB} = N_{TQ} \cdot p_{event} \qquad (6.17)$$

The cost function for this case is given by

$$K_{TQ}^{TD,RB}(N) = N \cdot [(1 + N_{TQ}p_{event})K_{sim} + p_{event}K_{RB}] + \frac{N}{N_{TQ}}(K_{sync} + K_{save}) \qquad (6.18)$$

Due to the term $\frac{1}{N_{TQ}} + N_{TQ}$, the structure of this formula suggests that there is some optimum length of the time quantum depending on the other parameters. To establish this optimum, the parameters must be measured in the specific simulation context.

6.5.4 Predictive Temporal Decoupling

Finally, we develop a cost function to compare the proposed PTD simulation approach with the strategies discussed in the preceding subsections. Since PTD changes the size of the TQ adaptively, it is not convenient to derive the cost function using time quanta. Therefore, we relate the simulation effort to a *time segment* (TS), which is equal to the maximum possible TQ.

Similar to the examination of temporally decoupled simulations with rollback mechanism, the cost function is derived in two steps. We assume that the TQ estimation algorithm has settled, so that the predicted TQ values and the corresponding error estimate are correct.

First, consider a time segment in which no external event occurs. Since this segment is simulated using TD, the associated simulation cost derived in Sect. 6.5.2—with N_{TS} being the length of the time segment in cycles—is given by

$$K_{TS,1}^{PTD} = K_{TQ}^{TD} = N_{TS} \cdot K_{sim} + K_{sync} \qquad (6.19)$$

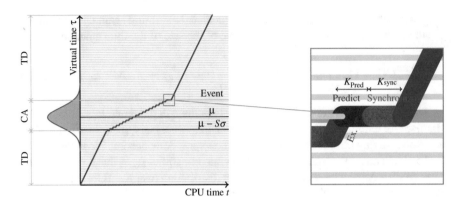

Fig. 6.13 PTD simulation of a model where K_{Pred} denotes the average prediction effort for a single event

Next, a segment with an event is investigated as shown in Fig. 6.13. The mean event interval is denoted by μ. The average number of cycles that have to be simulated in CA mode is given by $S\sigma$. Thus, the simulation cost for this segment consists of $S\sigma$ cycles simulated in CA mode plus the previous cycles simulated in TD mode.

To include the effort needed to predict the next position of the event, we add K_{Pred}.

$$K_{TS,2}^{PTD} = N_{TS} \cdot K_{sim} + (S\sigma + 1) \cdot K_{sync} + K_{Pred} \qquad (6.20)$$

Lastly, the probability that an event occurs in the observed time segment is introduced by $p_{event,TS}$. Therefore, the expected simulation cost can be expressed as

$$K_{TS}^{PTD} = (1 - p_{event,TS}) \cdot K_{TS,1}^{PTD} + p_{event,TS} \cdot K_{TS,2}^{PTD} \qquad (6.21)$$

Expanding and simplifying this term yields

$$K_{TS}^{PTD} = N_{TS} \cdot K_{sim} + (1 + S\sigma \cdot p_{event,TS})K_{sync} + p_{event,TS} \cdot K_{Pred} \qquad (6.22)$$

Hence, the total simulation cost for the simulation of N cycles is

$$K_{total}^{PTD}(N) = N \cdot K_{sim} + \frac{N}{N_{TS}} \left[(1 + S\sigma \cdot p_{event,TS}) \cdot K_{sync} + p_{event,TS} \cdot K_{Pred} \right] \qquad (6.23)$$

Assuming that the events are uniformly distributed over the observed time segment, the probability $p_{event,TS}$ can be expressed as

$$p_{event,TS} = N_{TS} \cdot p_{event,cycle} \qquad (6.24)$$

This imposes the further assumption that the probability that an event occurs in the observed time segment must not exceed one, i.e. only one event may occur during one time segment.

Moreover, it is possible to provide an expression for a combination of PTD and rollback mechanisms, assuming that the prediction error follows a gaussian distribution. Under this condition, the probability of a missed event can be expressed using the Q-function [12]:

$$Q(x) = \frac{1}{\sqrt{2\pi}} \int_x^\infty e^{-\frac{1}{2} \cdot u^2} du \qquad (6.25)$$

Since the simulation returns to CA mode as soon as the beginning of the safety margin is reached, an event which occurs after this point is caught exactly. Therefore, the probability of a rollback required after missing an event can be approximated by

$$p_{RB} = Q\left(\frac{S\sigma}{\sigma}\right) = Q(S) \qquad (6.26)$$

Eq. (6.26) does not take into account that an event must not occur before the previous synchronization. However, we can assume that the probability for this case is negligibly small. Furthermore, it is expected that an event occurs—in the sense of an average—in the middle of the observed time segment. Therefore, if an event occurs and a rollback is executed at the first synchronization after the TD mode, $(N_{TS} - S\sigma)$ cycles have to be resimulated.

The cost function for a segment with rollback, but not including the state-saving cost, is given by

$$K_{TS,3}^{PTD,RB} = (2N_{TS} - S\sigma) \cdot K_{sim} + 2K_{sync} + K_{Pred} + K_{RB} \qquad (6.27)$$

This term is combined with the ones derived before in Eqs. (6.19) and (6.20) yielding an expression for the simulation cost of PTD simulations with rollback.

We assume that, for a single time segment, the system state is saved at the beginning, incurring cost K_{save}.

$$
\begin{aligned}
K_{TS}^{PTD,RB} =& (1 - p_{event,TS}) K_{TS1}^{PTD} \\
& + p_{event,TS} \cdot \left[(1 - p_{RB}) K_{TS,2}^{PTD} + p_{RB} K_{TS,3}^{PTD,RB} \right] + K_{save}
\end{aligned}
\qquad (6.28)
$$

A simplified equation is obtained by replacing all known terms:

$$
\begin{aligned}
K_{TS}^{PTD,RB} =& N_{TS} \cdot K_{sim} + K_{save} + K_{sync} + \\
& p_{event,TS} \{ K_{pred} + (1 - p_{RB}) \cdot S\sigma \cdot K_{sync} + \\
& p_{RB} [(N_{TS} - S\sigma) \cdot K_{sim} + K_{RB}] \}
\end{aligned}
\qquad (6.29)
$$

Substituting p_{RB} with the value obtained via the Q-function yields

$$K_{TS}^{PTD,RB} = N_{TS} \cdot K_{sim} + K_{save} + K_{sync} +$$

$$p_{event,TS} \{ K_{pred} + Q(-S) \cdot S\sigma \cdot K_{sync} + \tag{6.30}$$

$$Q(S) [(N_{TS} - S\sigma) \cdot K_{sim} + K_{RB}] \}$$

For a simulation period of N cycles, assuming N to be an integer multiple of N_{TS}, the total simulation cost is

$$K_{total}^{PTD,RB}(N) = N \cdot K_{sim} + \frac{N}{N_{TS}} \left(K_{sync} + K_{save} \right)$$

$$\frac{N}{N_{TS}} p_{event,TS} \{ K_{pred} + Q(-S) \cdot S\sigma \cdot K_{sync} \tag{6.31}$$

$$Q(S) [(N_{TS} - S\sigma) \cdot K_{sim} + K_{RB}] \}$$

6.5.5 Discussion

The simulation cost models developed in this section provide a base for comparing the various simulation strategies systematically. Even though the parameter values (Table 6.1) are not known yet, the functions can be discussed qualitatively.

In Fig. 6.14 the cost functions are plotted for the fixed set of parameters listed in Table 6.2. The results for temporally decoupled simulations are similar to the ones shown in Sect. 6.6 and those obtained by Nitsche et al. [2] and Damm et al. [4]. Clearly, the simulation performance converges to a fixed limit for large TQ. For temporally decoupled simulation with rollback, an optimum TQ can be determined because the simulation cost may rise with larger TQ and correspondingly increasing rollback and resimulation overhead.

Not surprisingly, the best simulation performance is obtained using pure TD simulation, but this performance advantage comes at the cost of the worst timing accuracy among the compared strategies. Our new PTD approach regains this accuracy by automatically choosing an optimal TQ using a computationally cheap

Table 6.1 Summary of introduced cost function parameters

Parameter	Definition
K_{sim}	Average CPU time needed to simulate a single clock cycle
K_{sync}	Average CPU time needed to synchronize the simulation
K_{save}	Average CPU time needed to save the system state
K_{RB}	Average CPU time needed to restore previous system state
$K_{predict}$	Average CPU time needed for the prediction algorithm per cycle
p_{event}	Probability of event in regarded time interval

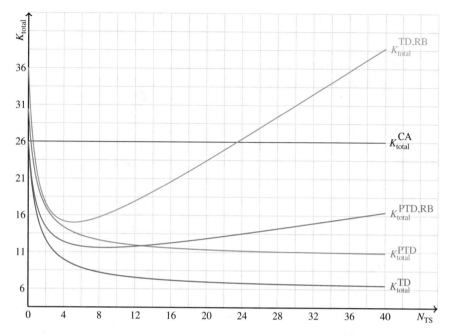

Fig. 6.14 Plot of the derived cost functions for fixed parameter values given in Table 6.2

Table 6.2 Parameter values used in Fig. 6.14

Parameter	Value
N	1
K_{sim}	1
K_{sync}	1
K_{RB}	1
K_{save}	0.5
p_{event}	0.02
K_{predict}	5
$Q(S)$	0.5

prediction algorithm and temporarily switching back to cycle-accurate simulation when required to capture an external event. This raises overall simulation cost again, but the structural performance advantage over CA simulation remains. Even in combination with a rollback algorithm, PTD reduces the overall simulation effort by reducing the number of rollback and resimulation steps.

The unknown parameters summarized in Table 6.1 are defined as averages because they may vary strongly over a complete simulation period. For example, a system could have a fast-simulating idle state and an expensive working state with different associated cost factors. To measure the average parameters, a simulation with different time quanta has to be done to measure several points of the cost function.

If a system contains multiple clocks, the cost functions for each clock domain can be converted and merged into a common cost function with reference to the cycle length of a selected master clock. A change of this reference clock can be modelled by multiplying the parameters with a constant value.

Our cost functions are generally applicable to simulations involving a single temporally decoupled module. However, system simulations may contain several such modules. It is yet unknown how the individual cost factors of each module can be combined to calculate the overall simulation cost. From an intuitive point of view, it can be expected that the simulation effort K_{sim} should be the sum of the values contributed by each block, but there is no formalism yet to calculate the combination overhead.

6.6   Simulation Results

To validate our approach under realistic conditions, we abstracted the simulation model of a smart-sensor system in order to quantify the effects of PTD. In our example system, illustrated in Fig. 6.1, the microcontroller receives input continually from a data source outside of the SystemC simulator.

To measure the effect of PTD separately from other possible influences on mixed-signal system simulation performance, we use a pure SystemC model for this experiment. This model is composed of two main modules, a microcontroller ISS and a stochastic event source. The latter represents an abstract behavioral equivalent of the analog circuit with regard to the externally generated input events. For simplicity, we do not use rollback mechanisms and accept any residual event processing latency in this experiment.

As explained previously, the event source generates a quasi-periodic sequence of events with random timing variation following a gaussian distribution. The event interval in this experiment is varied about a mean μ of 300 clock cycles with a standard deviation σ of 50 clock cycles. The model is simulated over a total virtual time of 5000 ms for a model clock period of $T_{clk} = 100$ ns. After each period of 1000 ms, the simulation method is changed, starting with CA simulation and stepping through TD and PTD modes with increasing safety margin parameter S. Since we do not use simulator coupling, no effort must be spent on analog simulation; thus, simulation performance depends only on the execution of the digital part of the model.

The effects of PTD can be seen in the simulation results shown in Fig. 6.15. The graph in Fig. 6.15a displays vertical short-term histograms of 500 measured consecutive event processing delays depending on the current simulation mode (CA, TD, or PTD) and the short-term average size of the time quantum (Fig. 6.15b). The color of a histogram point represents the corresponding bin count on a relative $\log_{10}$ scale shown on the right. The graph in Fig. 6.15d shows the resulting simulation performance for the selected simulation mode in model cycles executed per CPU second. To quantify the timing accuracy at TQ boundaries, the relative number of

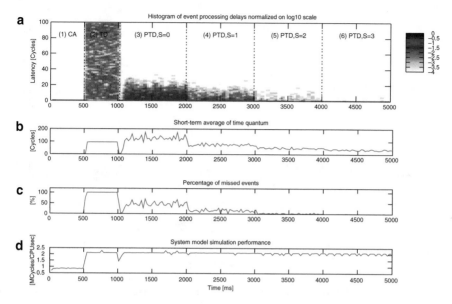

Fig. 6.15 Simulation of the system depicted in Fig. 6.1. In the most upper chart (**a**) we plotted normalized an top-viewed array of short time histograms of the delays δ using a logarithmic scale. The height of each histogram bar is color-coded for visualization. The other diagrams (**b–d**) show short time averages of the measured quantities

missed events is displayed in Fig. 6.15c. Note that the absolute performances and the corresponding performance ratios are model-dependent and cannot be generalized from this experiment.

In phase (1) the simulator operates in cycle-accurate mode. As can be expected, there is no event processing latency, but the simulation performance is low. Phase (2) is simulated using cycle-count-accurate TD [2] with a fixed value for the global time quantum. The resulting timing error is approximately uniformly distributed over the whole TQ because of uncorrelated TQ and event intervals.

Our proposed PTD method that predicts the optimum size and position of the TQ is applied in the phase (3) of the simulation. In this experiment, the predictor is forced to start without initial parameter estimates. Thus, a short startup period with slow, cycle-accurate simulation is required to establish the predictor parameters. Subsequently, the time quantum increases to the maximum possible value while simultaneously reducing the corresponding timing error. Simulation performance remains as high as in phase (2) even though the prediction algorithm is active. According to Sect. 6.4, we can expect to capture approximately 50% of the events accurately because simulation switches to CA mode at the point in time predicted for the next event. This simulation results in the time interval from 1000 to 2000 ms are in line with this expectation.

In the following phases (4)–(6), the variance of the prediction error is used to determine the safety margin $S\sigma$ required to reduce event processing latency to a

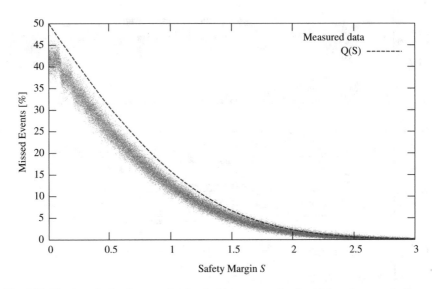

Fig. 6.16 Number of missed events simulated with respect to the size of the safety margin S

given statistical acceptance level. Although the time quantum is reduced as S is varied from 1 to 3, the impact on overall simulation performance is low while the set of missed events can reduced to a few statistical outliers.

The influence of the safety margin parameter S on the proportion of missed events has been examined in Fig. 6.16. The measured data is slightly below the predicted curve due to pessimistic rounding of the predicted event times to clock cycle boundaries. For normally distributed event distances, this proves the possibility to control the number of missed events via the size of the safety margin.

Finally, we apply the cost models from Sect. 6.5 to measure the simulation cost in terms of CPU effort. The simulation results are shown in Fig. 6.17; they were generated on the platform described in Table 6.3. For each TQ, several measurement points were generated to eliminate the influence of caching or other non-deterministic effects in the operating system of the simulation host. The displayed curves for TD and PTD simulation differ by the expected small offset caused by the additional CPU resources required for the prediction algorithm and the short intermediate cycle-accurate simulation phases in PTD mode. The numerical values of the measured parameters are listed in Table 6.4.

The simulated data shows that our proposed method combines the high performance of TD simulation with a timing accuracy close to cycle-accurate simulation.

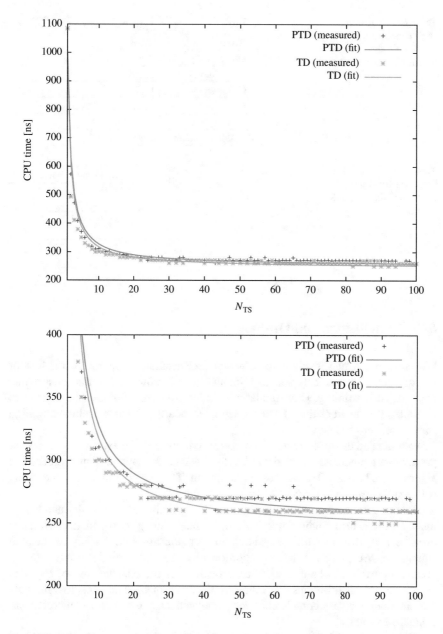

Fig. 6.17 Measurement of the mean simulation cost for a single simulated clock cycle for a microcontroller ISS with temporally decoupled command interpretation thread

Table 6.3 Hardware environment used for simulation performance measurements

CPU	AMD Phenom(tm) II X6 1090T
RAM	16GB
Operating system	CentOS 5.9
Linux kernel	2.6.18
GCC version	4.1.2
GCC target	i686-pc-linux-gnu
GCC options	-O3
SystemC version	2.2.0, TLM 2.0.1

Table 6.4 Estimated cost function parameters for decoupling the main ISS process

Parameter	Value
$K_{\text{sim}}^{\text{TD}}$	$244.7 \frac{ns}{\text{cycle}}$
$K_{\text{sim}}^{\text{PTD}}$	$251.8 \frac{ns}{\text{cycle}}$
K_{sync}	$722.4 \frac{ns}{\text{sync}}$
$p_{\text{event}} K_{\text{Pred}}$	$33.4 \frac{ns}{\text{cycle}}$

6.7 Conclusions and Outlook

We presented a new method for efficient and accurate temporal decoupling of heterogeneous system components without a priori knowledge of the subsystems' communication timing. Our simulation results prove that our method is able to minimize the inherent timing inaccuracy of standard TD while maintaining high simulation performance.

We modeled the time between subsequent communication events as a stochastic process and predicted these times using a standard adaptive signal processing algorithm. Thus, PTD estimates an optimum TQ, which minimizes the delay between the event and the next synchronization cycle.

For an assumed statistical distribution, it is possible to reduce the number of missed events to a minimum by adding a safety margin around the estimated event arrival time, in which the simulation temporarily returns to cycle-accurate simulation. We proved that the proportion of events that are still missed due to prediction errors can be quantified analytically. For the realistic case of bounded event interval distributions, we showed that our method is able to eliminate the simulation errors completely. The simulated data matches the theoretically calculated values.

We applied our PTD method experimentally to the simulation of the virtual prototype of a smart-sensor system which includes a temporally decoupled micro-controller ISS for firmware execution. In this example, we showed that it is possible to combine the performance of a TD simulation with the accuracy of a purely cycle-accurate simulation and provide an estimate for remaining event processing latencies. Other fields of application could be approximate modeling of real-time systems and other scheduling problems with asynchronous non-deterministic events.

Future work could help to improve PTD through better prediction and estimation algorithms for a wider class of problems. This applies in particular to non-stationary event sources and to higher model orders, which would be required for dealing efficiently with correlated event bursts. Moreover, the presented methods could be extended to predictive temporal decoupling of several system components in the same simulation model.

Acknowledgements This work has been carried out in the research project ANCONA, funded by the German Federal Ministry of Education and Research (BMBF) in the ICT2020 program under grant no. 16ES021. The results are based on previous work from the project *GreenSense*, funded by the Thuringian Ministry of Economics, Labor and Technology (TMWAT) and the European Social Fund (ESF) under grant no. 2011 FGR 0121.

References

1. IEEE Standard Association. IEEE Standard for Standard SystemC Language Reference Manual. IEEE, New York (2012). DOI 10.1109/IEEESTD.2012.6134619
2. Nitsche, G., Glaeser, G., Nuernbergk, D., Hennig, E.: In: Brandt, J., Schneider, K. (eds.) Methoden und Beschreibungssprachen zur Modellierung und Verifikation von Schaltungen und Systemen (MBMV), pp. 121–132. Verlag Dr. Kovac, Kaiserslautern (2012)
3. Monton, M., Engblom, J., Burton, M.: IEEE Trans. Very Large Scale Integr. VLSI Syst. **PP**(99), 1 (2012). DOI 10.1109/TVLSI.2011.2181881
4. Damm, M., Grimm, C., Haase, J., Herrholz, Nebel, W.: In: Forum on Specification, Verification and Design Languages, 2008. FDL 2008, pp. 25–30 (2008). DOI 10.1109/FDL.2008.4641416
5. Gladigau, J., Haubelt, C., Teich, J.: IEEE Trans. Comput. Aided Des. Integr. Circ. Syst. **31**(10), 1572 (2012). DOI 10.1109/TCAD.2012.2205148
6. Razaghi, P., Gerstlauer, A.: IEEE Embed. Syst. Lett. **4**(1), 5 (2012). DOI 10.1109/LES.2012.2186281
7. Salimi Khaligh, R., Radetzki, M.: In: Design, Automation Test in Europe Conference Exhibition (DATE), 2010, pp. 1183–1188 (2010)
8. Schirner, G., Domer, R.: IEEE Trans. Comput. Aided Des. Integr. Circ. Syst. **26**(9), 1688 (2007). DOI 10.1109/TCAD.2007.895757
9. Bringmann, O., Ecker, W., Gerstlauer, A., Goyal, A., Mueller-Gritschneder, D., Sasidharan, P., Singh, S.: In: Design, Automation Test in Europe Conference Exhibition (DATE), 2015, pp. 1698–1707 (2015)
10. Brandner, F.: In: Proceedings of the 12th International Workshop on Software and Compilers for Embedded Systems, SCOPES '09, pp. 71–80. ACM, New York, NY (2009)
11. Haykin, S.: Adaptive Filter Theory, 3rd edn. Prentice-Hall, Upper Saddle River, NJ (1996)
12. Moon, T., Stirling, W.: Mathematical Methods and Algorithms for Signal Processing. Prentice Hall, Upper Saddle River (2000)
13. Lin, Y.B., Preiss, B.R., Loucks, W.M., Lazowska, E.D.: In: Proceedings of the 7th Workshop on Parallel and Distributed Simulation, pp. 3–10. IEEE Computer Society Press, Los Alamitos (2001)

Part IV
AMS Circuits and Systems

Chapter 7
SystemC-AMS Simulation of Conservative Behavioral Descriptions

Sara Vinco, Michele Lora, and Mark Zwolinski

7.1 Introduction

SystemC has long been considered the reference language for electronic system-level design, as it supports both hardware and software and the integration of multiple levels of abstraction, including Register Transfer Level (RTL) and transactional level [1]. However, the increasing presence, in embedded systems, of analog components and Micro Electro-Mechanical Systems (MEMS) limits the generality of SystemC [2]. Indeed, these types of component require the support of continuous time and conservative behaviors, which cannot be modeled with a discrete event simulator.

In response to this, Accellera has standardized the SystemC-AMS extension [3]. SystemC-AMS provides a number of predefined levels of abstraction that reproduce linear continuous time models with different degrees of accuracy and adherence to physical behaviors. Unfortunately, such abstraction levels (briefly outlined in Fig. 7.1) do not model all types of analog models. The *Linear Signal Flow* (LSF) level of abstraction focuses on behavioral, continuous time systems, but it does not support the modeling of conservative systems. On the other hand, *Electrical Linear Network* (ELN) is conservative, but it does not support behavioral models.

S. Vinco (✉)
Politecnico di Torino, Turin, Italy
e-mail: sara.vinco@polito.it

M. Lora
Università di Verona, Verona, Italy
e-mail: michele.lora@univr.it

M. Zwolinski
University of Southampton, Southampton, UK
e-mail: mz@ecs.soton.ac.uk

© Springer International Publishing Switzerland 2016
R. Drechsler, R. Wille (eds.), *Languages, Design Methods, and Tools
for Electronic System Design*, Lecture Notes in Electrical Engineering 385,
DOI 10.1007/978-3-319-31723-6_7

Fig. 7.1 Design space covered by the proposed approach *w.r.t.* SystemC-AMS. The methodology bridges the gap between LSF and ELN, by targeting behavioral and conservative descriptions (*ABM*)

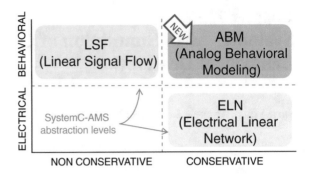

Furthermore, SystemC-AMS does not yet support non-linear modeling [4] and it can not therefore be considered a replacement for SPICE [5] or Verilog-AMS [6].

The resulting gap between ELN and LSF thus misses descriptions that are both behavioral and conservative, and that are commonly used for the modeling of MEMS and other analog components [7, 8]. The limited flexibility of SystemC-AMS forces designers to adopt other AMS HDLs (e.g., Verilog-AMS) for modeling these kinds of components, thus reducing the applicability of SystemC-AMS.

The key idea of this work is to bridge the gap between LSF and ELN, to represent models that are both behavioral and conservative in SystemC-AMS. This new level of abstraction, called *Analog Behavioral Modeling* (ABM), is demonstrated with a methodology that exploits existing SystemC-AMS constructs. The goal is to show that SystemC, with its AMS extensions, can be used as a general embedded system modeling and simulation framework even in the presence of analog circuitry and MEMS. It is important to note that this work does not define new SystemC-AMS libraries, but it rather uses ELN primitives in an innovative way.

The main contributions of this work are:

- *identification of the ABM level of abstraction*, necessary for overall embedded system simulation in a SystemC-based environment;
- definition of a *sound methodology* for modeling ABM components in SystemC-AMS, by using existing primitives in a novel way;
- *validation of the ABM level against Verilog-AMS*, to show that those Verilog-AMS descriptions that do not fall into the ELN and LSF domains can now be correctly represented in SystemC-AMS;
- *automation of the proposed methodology* by automatically converting Verilog-AMS models into ABM SystemC-AMS models. This simplifies the application of the methodology to complex industrial case studies.

As a side effect of the proposed methodology, Verilog-AMS models can be automatically converted into SystemC-AMS code for easy integration into a virtual platform including analog models. This avoids the use of CPU intensive co-simulation frameworks, thus noticeably speeding up the simulation of a virtual platform.

The paper is organized as follows. Section 7.2 provides the necessary background on Verilog-AMS and SystemC-AMS. Sections 7.3 and 7.4 focus on the proposed methodology. Finally, Sect. 7.5 applies the proposed approach to an industrial case study and Sect. 7.6 draws some conclusions.

7.2 Background

This Section provides the necessary background on the adopted languages for analog and mixed signal modeling: Verilog-AMS (Sect. 7.2.1) and SystemC-AMS (Sect. 7.2.1).

7.2.1 Verilog-AMS

Verilog-AMS is one of the most widely used languages for analog and continuous time modeling [6]. The system solver is essentially the same as that used in SPICE [5]. A circuit is modeled in terms of an abstract graph of nodes (that can also be used for external connectivity) connected by branches [9]. The system state is defined in terms of voltages ($V()$) and currents ($I()$) associated with nodes and branches. The numerical values of potential differences and currents can be used in expressions with the access functions $V()$ and $I()$. Relationships between nodes are modeled with differential algebraic equations (DAEs), written as *simultaneous statements*. The *contribution operator* $<+$ models a simultaneous statement summing multiple contributions to the branch current (or voltage) as a function of other branch voltages and currents.

Conservative modeling is imposed by the requirement that the sum of currents leaving any node must be equal to zero at any time (thus reflecting Kirchhoff's Current Law). This condition is managed by the internal solver of the Verilog-AMS simulator, and thus must not be explicitly modeled by the designer.

The simulator internal solver uses the simultaneous statements and conservative conditions to build a system matrix. Numerical integration methods are used to solve the system of DAEs. Continuous time is modeled as a sequence of discrete time points, such that the time step is optimized to minimize errors while maximizing efficiency.

Non-linear equations are solved iteratively at each discrete time step to determine the system state over time. Typically, the Newton-Raphson method [10] is used for linearization.

7.2.2 SystemC-AMS

SystemC-AMS is the extension of the SystemC framework for modeling analog and mixed-signal systems [3]. Its role is to provide a higher level view of mixed-signal and analog systems, to allow early simulation and validation of the overall system. For this reason, SystemC-AMS supports only linear and time-invariant descriptions, and is incapable of solving non-linear functions [4].

To cover a wide variety of domains, SystemC-AMS provides three different abstraction levels, supporting different communication styles and representations with respect to the physical domain:

- *Timed Data-Flow* (TDF) models are scheduled statically by considering their producer-consumer dependencies in the discrete time domain;
- *Linear Signal Flow* (LSF) supports the modeling of continuous time through a library of pre-defined primitive modules (e.g., integration, delay), each associated with a linear equation;
- *Electrical Linear Network* (ELN) level models electrical networks through the instantiation of predefined primitives, e.g., resistors or capacitors, where each primitive is associated with electrical equations.

A *SystemC-AMS internal solver* analyses the ELN and LSF components to derive the equations modeling system behavior, that are solved to determine the system state at any simulation time.

The main difference between LSF and ELN is in the adherence to physical laws. LSF is *non-conservative* and it expresses behaviors as directed flows of continuous-time signals or quantities. On the other hand, ELN is *conservative*, i.e., the derived set of equations is extended by the internal solver to satisfy the conservation laws (Kirchhoff's laws).

7.3 Methodology Overview

The goal of the proposed approach is to prove that conservative and behavioral descriptions can be modeled in SystemC-AMS. To this extent, the starting point of the methodology is a Verilog-AMS behavioral description, made up of a set of simultaneous statements that assemble voltages or currents to describe the state of the electrical circuit nodes. Due to the limitations of SystemC-AMS, the models are strictly linear and time-invariant.

The standard Verilog-AMS simulation flow is depicted on the left-hand side of Fig. 7.2. The Verilog-AMS internal solver takes the simultaneous statements as the input and derives both the user defined equations and the conservative ones. The resulting equation set is used to build the numerical matrices that determine the system state.

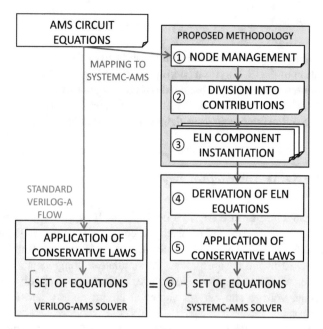

Fig. 7.2 Overview of the proposed methodology

The methodology to convert the Verilog-AMS code into SystemC-AMS is based on reproducing the final equation set in the SystemC-AMS environment, through the flow depicted on the right-hand side of Fig. 7.2. First of all, Verilog-AMS nodes are mapped to SystemC-AMS nodes (①), and Verilog-AMS simultaneous statements are divided into basic contributions (②). Then, each contribution is mapped to a basic SystemC-AMS ELN element, where the equation associated with each ELN module is the same as the original Verilog-AMS contribution (③). The methodology determines how to connect the ELN modules (i.e., in parallel or in series), so that the bindings describe the same relationship between voltages and currents as in the original Verilog-AMS simultaneous statement. The ELN system is then managed by the SystemC-AMS internal solver, that builds the corresponding equations (④) and adds conservative laws (⑤). The resulting equation system will thus reflect the Verilog-AMS one (⑥).

The choice of the ELN model of computation allows us to delegate the application of conservation laws to the internal solver. This is an important feature, as adding conservative laws implies reconstructing the circuit topology from the AMS equations, which can be far from trivial.

It is important to note that both the basis of the methodology and the correctness of the proposed approach lie in the construction of the same equation set, that is then solved in the same way by the Verilog-AMS and SystemC-AMS solvers.

7.4   Methodology

The following sections describe the methodology in detail. This work focuses on the construction of the ELN system (steps ① to ③ in Fig. 7.2). The remaining steps (i.e., the bottom box on the right-hand side of Fig. 7.2) are automatically performed by the SystemC-AMS internal solver. The visual representation of ELN modules adopted in the following figures is as defined by the SystemC-AMS standard [3].

7.4.1   The ABM Abstraction Level

The ABM abstraction level comprises descriptions mixing characteristics typical of digital behavioral models and of electrical conservative ones. ABM models are *behavioral* in that they do not directly reflect a hardware or circuit implementation, but they are used in the design process to simulate a component's behavior. At the same time, ABM models are *conservative* as they adopt circuit elements and constructs (e.g., voltage and current values at circuit nodes), and thus abide by conservation laws.

These characteristics do not fit in any of the SystemC-AMS abstraction levels. Nonetheless, they are widely supported by other AMS HDLs for the design of components such as MEMS and analog circuitry [7, 8]. It is thus necessary to extend SystemC-AMS, to improve its coverage and effectiveness. To avoid the burden of implementing a new SystemC-AMS abstraction level (and thus new classes and libraries), this work proposes a methodology that converts ABM descriptions, modeled in other AMS HDLs, to SystemC-AMS ELN constructs. This guarantees the correctness of the underlying solution techniques, and it preserves compatibility with any SystemC-AMS description.

7.4.2   ELN Terminology

ELN descriptions are based on the instantiation of an electrical network composed of electrical primitives (i.e., *ELN modules*), connected together at *electrical nodes*. Each ELN module contributes to the equation system with a particular set of equations defining the mathematical relations at each node of the network.

The interface of an electrical primitive is composed of a set of *terminals*. A negative (n)terminal and a positive (p) terminal are present on the interface of basic passive (linear) electrical components (i.e., capacitor, inductor and resistor) and of voltage or current sources (left-hand side of Fig. 7.3).

A particular sub-set of primitives, largely employed by the proposed methodology, is that of *controlled sources*. Controlled source primitives determine a current or voltage value on an output branch, whose generation linearly depends on the

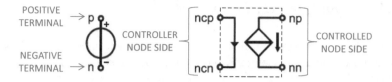

Fig. 7.3 ELN terminology applied to an independent source (*left*) and to a controlled source (*right*)

current or voltage value of an input branch. For this reason, controlled sources present two different interfaces, as depicted on the right-hand side of Fig. 7.3: a *control* interface and a *controlled* interface. Each interface represents an *electrical branch*, i.e., they are composed of a negative terminal and a positive terminal (i.e., ncn and ncp terminals for the control interface, and nn and np for the controlled interface). Thus, the controlled sources introduce a set of relations where the value of voltage or current on the controlled branch is proportional to the value of voltage or current on the control branch.

7.4.3 Circuit Node Management

The first declaration added to the SystemC-AMS code is the instantiation of ground, declared as a node of type sca_node_ref. Verilog-AMS nodes are mapped to SystemC-AMS ELN circuit nodes (of type sca_node). Each node is then connected to ground through a 1 GΩ resistor, by using the ELN sca_r primitive. This is identical to the Gmin conductance that SPICE automatically inserts between each node and ground, and it helps to ensure the solution of the equation system specified by the circuit.

7.4.4 Division into Contributions

SystemC-AMS is less expressive than Verilog-AMS, i.e., it supports a more restricted range of constructs and ELN models can be composed only of instances of the predefined primitives [8, 10]. Furthermore, SystemC-AMS does not allow this set of predefined primitives to be extended. E.g., in SystemC-AMS a voltage value can be controlled only by one voltage or current contribution, while Verilog-AMS allows any number of contributions. Thus, a general Verilog-AMS simultaneous statement must be reproduced by connecting a number of ELN elements.

Given a Verilog-AMS description, our technique identifies the contributions comprising each simultaneous statement by finding the largest sub-equation that can be represented by a single ELN object. In linear and time-invariant descriptions this corresponds to breaking the equation into the single addends.

7.4.5 Mapping to ELN Components

The remainder of this section shows how a set of template equations is mapped to ELN primitives to model their individual contributions and how such primitives are connected.

7.4.5.1 Voltage Sources

Voltage source Verilog-AMS equations use a number of contributions to assign a voltage level to a circuit node. Contributions can be of three main types: independent, voltage-controlled and current-controlled. A complete example of a voltage source equation is shown in Fig. 7.4.

Independent voltage sources assign a numerical voltage value, and they correspond to contributions like:

$$V(a) \ <+ \ 8.01$$

(i.e., contribution 3 in Fig. 7.4). They are implemented by using a sca_vsource ELN module, where the voltage value is an instantiation parameter (i.e., +8.01). The module interface has only a positive terminal, connected to the controlled node (a), and a negative terminal, connected to ground (gnd).

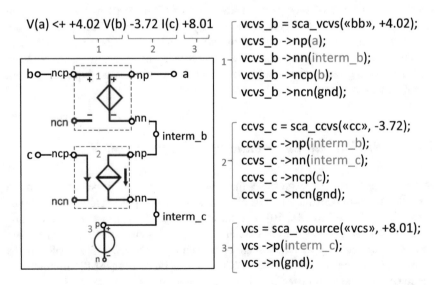

Fig. 7.4 Example of voltage source equation (*top left*) with the corresponding SystemC-AMS code (*right*) and ELN module connection (*bottom left*). Non-connected terminals are connected to ground

A *voltage controlled voltage source* is a voltage source whose value depends on the voltage between a pair of circuit nodes. An example is contribution 1:

$$V(a) \; <+ \; +4.02 \; V(b)$$

This is implemented by using the `sca_vcvs` ELN module, where the scaling factor is an instantiation parameter (i.e., +4.02). The module interface has a controlling node side (whose positive terminal is connected to b) and a controlled node side (whose positive terminal is connected to a). The negative controlling terminal is connected to ground.

A *current controlled voltage source* describes a voltage source whose value depends on the current through a circuit branch. An example is contribution 2:

$$V(a) \; <+ \; -3.72 \; I(c).$$

Such contributions are implemented by using the `sca_ccvs` ELN module, connected to node a as the controlled node and to node c as the controlling node.

If a Verilog-AMS voltage source equation is made up of more than one contribution, SystemC-AMS instances are *connected in series*. This is achieved by creating intermediate nodes that connect the nn terminal of a primitive with the np terminal of the next primitive. In this way, voltage values add up and any new contribution is added in series with the former ones. In Fig. 7.4, this is achieved by introducing intermediate nodes `interm_b` (that connects contributions 1 and 2) and `interm_c` (that connects contributions 2 and 3).

7.4.5.2 Current Sources

Current source Verilog-AMS equations are the complement of voltage source equations, i.e., they use a number of contributions to assign an input current to a circuit node. A complete example is shown in Fig. 7.5.

An *independent current source* assigns a numerical current value and is implemented by using the `sca_isource` ELN module (contribution 3 in Fig. 7.5). A *voltage controlled current source* defines a current source whose value depends on the voltage level at a certain circuit node (contribution 1 in Fig. 7.5). These kinds of contributions are mapped to `sca_vccs` ELN modules. Finally, a *current controlled current source* describes a current source whose value depends on the current flowing through a certain circuit branch (contribution 2 in Fig. 7.5). Such contributions are implemented by using `sca_cccs` ELN modules.

If a Verilog-AMS current source equation is made up of more than one contribution, SystemC-AMS instances are *connected in parallel*. The ncp terminal of each module is connected to the controlling node (b and c) and the np (or p) terminal is connected to the controlled node (a). In this way, the voltage is the same across all involved circuit branches and the current is summed at the controlled node a.

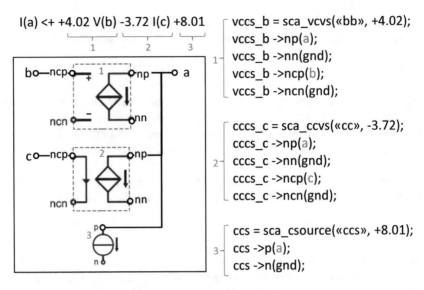

I(a) <+ +4.02 V(b) -3.72 I(c) +8.01

```
vccs_b = sca_vcvs(«bb», +4.02);
vccs_b ->np(a);
1- vccs_b ->nn(gnd);
vccs_b ->ncp(b);
vccs_b ->ncn(gnd);

cccs_c = sca_ccvs(«cc», -3.72);
cccs_c ->np(a);
2- cccs_c ->nn(gnd);
cccs_c ->ncp(c);
cccs_c ->ncn(gnd);

ccs = sca_csource(«ccs», +8.01);
3- ccs ->p(a);
ccs ->n(gnd);
```

Fig. 7.5 Example of current source equation (*top left*) with the corresponding SystemC-AMS code (*right*) and ELN module connection (*bottom left*). Non-connected terminals are connected to ground

7.4.5.3 Differential Constructs

Differential contributions are more complex than voltage or current ones, as they model a derivative (or integrative) relationship between the current or voltage of two separate circuit nodes. SystemC-AMS, however, restricts differential behaviors to dependencies on single network nodes, through the adoption of capacitors (sca_c ELN primitive) or inductors (sca_l ELN primitive). To overcome this limitation, it is necessary to introduce an intermediate node that has no physical correspondence in the circuit, but that is used for describing the differential dependence.

All the differential contributions are mapped using the generic topological pattern depicted in Fig. 7.6, where colors depict the physical quantities involved: light blue for current and red for voltage, while yellow portions are dependent on the type of contribution to reproduce.

Component 1 is a controlled current-source, as indicated by the controlled side in Fig. 7.6 (in blue). The control side (in yellow) depends on the modeled contribution: if the argument of the derivative construct is a voltage, component 1 is a voltage controlled current source; else, if the argument is a current, component 1 is a current controlled current source.

Component 3 is a voltage-controlled source, as indicated by the control side in Fig. 7.6 (in red). The controlled side (in yellow) reflects the target of the Verilog-AMS contribution statement: if the target is a voltage, component 3 is a voltage controlled voltage source; else if the target is a current, component 3 is a voltage controlled current source.

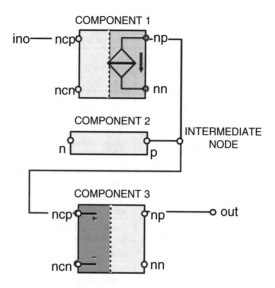

Fig. 7.6 Generic topological pattern used to implement differential contributions. All disconnected terminals are connected to ground

Component 2 (in yellow in Fig. 7.6) is used to create the differential relation between the current value controlled by component 1 and the voltage value controlling Component 3. The component is an inductor whenever the differential contribution is derivative, and it is a capacitor in the case of an integrative contribution. Given $I_{np,nn}$ the current flowing through terminals np and nn of Component 1, and $V_{ncp,ncn}$ the voltage on the branch between the terminals ncp and ncn of Component 3, the relationship described by Component 2 is thus:

$$V_{ncp,ncn} = \int I_{np,nn} dt$$

in the case of a derivative contribution (i.e., Component 2 is a capacitor), and

$$V_{ncp,ncn} = \frac{dI_{np,nn}}{dt}$$

in the case of an integrative contribution (i.e., Component 2 is an inductor).

Considering the derivative and the integrative operators of Verilog-AMS, we can restrict all possible configurations of the topological pattern to the eight cases summarized in Table 7.1. For each case, the table shows the SystemC-AMS primitives used to instantiate Components 1, 2 and 3. The remainder of this section shows the application to two example cases, i.e., a derivative contribution of type $I(out) < +kddt(V(in_1))$ and an integrative contribution of type $I(out) < +kidt(V(in_1))$, respectively.

Table 7.1 Summary of the components employed to map differential contributions

Contribution	Component 1	Component 2	Component 3
$I(out) <+ k\ ddt(I(in_1))$	Current controlled current source	Inductor	Voltage controlled current source
	sca_cccs	sca_l	sca_vccs
$I(out) <+ k\ ddt(V(in_1))$	Voltage controlled current source	Inductor	Voltage controlled current source
	sca_vccs	sca_l	sca_vccs
$V(out) <+ k\ ddt(I(in_1))$	Current controlled current source	Inductor	Voltage controlled voltage source
	sca_cccs	sca_l	sca_vcvs
$V(out) <+ k\ ddt(V(in_1))$	Voltage controlled current source	Inductor	Voltage controlled current source
	sca_vccs	sca_l	sca_vcvs
$I(out) <+ k\ idt(I(in_1))$	Current controlled current source	Capacitor	Voltage controlled current source
	sca_cccs	sca_c	sca_vccs
$I(out) <+ k\ idt(V(in_1))$	Voltage controlled current source	Capacitor	Voltage controlled current source
	sca_vccs	sca_c	sca_vccs
$V(out) <+ k\ idt(I(in_1))$	Current controlled current source	Capacitor	Voltage controlled voltage source
	sca_cccs	sca_c	sca_vcvs
$V(out) <+ k\ idt(V(in_1))$	Voltage controlled current source	Capacitor	Voltage controlled current source
	sca_vccs	sca_c	sca_vcvs

Derivative Contributions

Consider a derivative contribution of the form of the second entry of Table 7.1:

$$I(a) <+ ddt(+4.02\ V(b))$$

as exemplified in Fig. 7.7. Given the derivative nature of the contribution, the circuit requires a new node (interm), that is connected to an inductor (i.e., an instance of a sca_l ELN module). This adds the following equation (Eq. 2 in Fig. 7.7):

$$V(interm) = ddt(I(interm))$$

Then, two equations are necessary to bind the values in a and b to the current and voltage values in the new node interm. In Fig. 7.7, node a is modeled as a current source dependent on the voltage in interm (the dependency is implemented as an instance of sca_vccs). This adds equation 3:

$$I(a) = +4.02\ V(interm)$$

I(a) <+ ddt(+4.02 V(b))

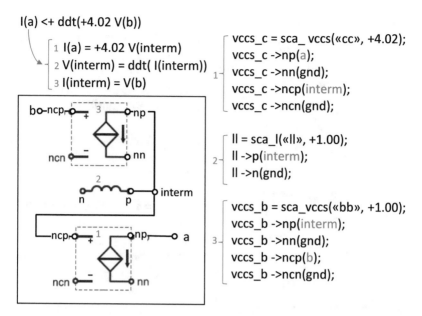

1 I(a) = +4.02 V(interm)
2 V(interm) = ddt(I(interm))
3 I(interm) = V(b)

1 — vccs_c = sca_ vccs(«cc», +4.02);
vccs_c ->np(a);
vccs_c ->nn(gnd);
vccs_c ->ncp(interm);
vccs_c ->ncn(gnd);

2 — II = sca_I(«II», +1.00);
II ->p(interm);
II ->n(gnd);

3 — vccs_b = sca_vccs(«bb», +1.00);
vccs_b ->np(interm);
vccs_b ->nn(gnd);
vccs_b ->ncp(b);
vccs_b ->ncn(gnd);

Fig. 7.7 Example of derivative equation with corresponding individual contributions and equations (*top left*), SystemC-AMS code (*bottom left*) and ELN module connection (*right*)

The current through `interm` is controlled by the voltage at b (the dependency is implemented as an instance of `sca_vccs`). This adds equation 1:

$$I(interm) = V(b)$$

The resulting system of equations will thus reconstruct the original dependency between nodes:

$$I(a) = +4.02V(interm) = +4.02ddt(I(interm))$$

$$= +4.02ddt(V(b))$$

This reflects the mapping defined in Table 7.1. The resulting SystemC-AMS subsystem is depicted on the left-hand side of Fig. 7.7. The sub-system can be further connected to other ELN models if it is included in more complex simultaneous statements, by adopting either parallel or series compositions.

Integrative Contributions

An integrative contribution illustrates the sixth case of Table 7.1:

$$I(a) <+ idt(+4.02 V(b))$$

I(a) <+ idt(+4.02 V(b))

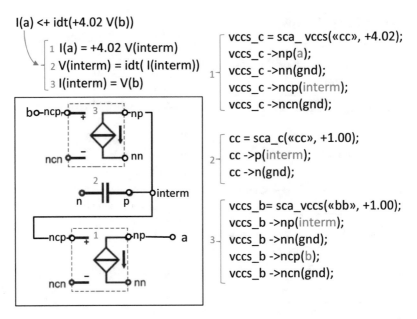

$$
\begin{aligned}
&1 \quad I(a) = +4.02 \ V(interm) \\
&2 \quad V(interm) = idt(\ I(interm)) \\
&3 \quad I(interm) = V(b)
\end{aligned}
$$

1—
```
vccs_c = sca_ vccs(«cc», +4.02);
vccs_c ->np(a);
vccs_c ->nn(gnd);
vccs_c ->ncp(interm);
vccs_c ->ncn(gnd);
```

2—
```
cc = sca_c(«cc», +1.00);
cc ->p(interm);
cc ->n(gnd);
```

3—
```
vccs_b= sca_vccs(«bb», +1.00);
vccs_b ->np(interm);
vccs_b ->nn(gnd);
vccs_b ->ncp(b);
vccs_b ->ncn(gnd);
```

Fig. 7.8 Example of integrative equation with corresponding individual contributions and equations (*top left*), SystemC-AMS code (*bottom left*) and ELN module connection (*right*)

as shown in Fig. 7.8. The circuit is extended with a new node (interm), used to represent the integrative dependency as a capacitor (an instance of the sca_c ELN module). This adds to the system equation 2:

$$
V(interm) \ = \ idt(I(interm))
$$

Then, two equations are necessary to bind the values in a and b to the current and voltage values in the new node interm, similarly to the solution proposed for derivative contributions:

$$
I(a) \ = \ +4.02 \ V(interm)
$$

$$
I(interm) \ = \ V(b)
$$

The resulting set of equations will thus reproduce the original dependency between nodes:

$$
I(a) \ = \ +4.02V(interm) = \ +4.02idt(\ I(interm))
$$

$$
= \ +4.02idt(\ V(b))
$$

The resulting SystemC-AMS sub-system is depicted on the left-hand side of Fig. 7.8. The sub-system can be further connected to other ELN models if it is included in more complex simultaneous statements, by adopting either parallel or series compositions.

7.5 Experimental Results

This section demonstrates the effectiveness of the proposed approach in terms of accuracy and simulation time. All experiments were evaluated on an i7 3.2GHz processor with 16GB RAM, running Ubuntu 14.04. Verilog-AMS descriptions have been simulated using Mentor's Questa 13.1 simulator [11].

7.5.1 Methodology Automation

Manual application of the proposed methodology is a tedious error-prone process, and application to industrial case studies could be extremely difficult. For this reason, we implemented an automatic tool *ABACuS* (Analogue BehAvioural Conservative Systemc-ams). *ABACuS* leverages the academic license version of HIFSuite to ease the conversion process [12]. Verilog-AMS descriptions are analyzed and translated into the HIFSuite internal format (HIF). The code generated at this point is a tree-structured XML-like representation of the original code. *ABACuS* applies a number of processing steps to the HIF description to automate the methodology, including contribution identification and construction of the ELN system. This leads to a new HIF description, containing the instantiation and connection of the corresponding ELN primitives. The HIF description is then converted to SystemC-AMS by means of the HIFSuite *hif2sc* back-end tool.

7.5.2 Methodology Validation

The first step to validate the propose methodology is to evaluate the accuracy of mapping single types of contribution, as detailed in Sect. 7.4. The accuracy is evaluated with 12 case studies, each targeting a single type of contribution. The case studies were implemented in Verilog-AMS and then converted to SystemC-AMS via *ABACuS*. The main characteristics of each case study are reported in Table 7.2, in terms of target contribution type and simulation time. Case studies are fed with sinusoidal inputs with 1KHz frequency, so that the outputs can be easily controlled and compared *w.r.t.* the expected theoretical results. In particular, the system of equations described using Verilog-AMS has been computed symbolically and solved for every time instant where a sample is collected by the SystemC-AMS execution. The SystemC-AMS simulation is run with an integration and sampling period of 10ns.

Simulation times in Table 7.2 refer to the amount of time needed to perform 1 s of transient simulation of the circuit implementing the given basic contribution. For all the cases depicted in Table 7.2, the time needed for simulating the initial

Table 7.2 Validation of the mapping of each type of contribution to ELN constructs

Case study	Target contribution	Simulation time (s)	Normalized RMSE
1	$V\,(out) <+\ k_1 V\,(in_1) + \ldots + k_n V\,(in_n) +c$	69.05	4.441e−7
2	$V\,(out) <+\ k_1 I\,(in_1) + \ldots + k_n I\,(in_n) +c$	70.59	4.441e−7
3	$I\,(out) <+\ k_1 V\,(in_1) + \ldots + k_n V\,(in_n) +c$	69.11	4.441e−7
4	$I\,(out) <+\ k_1 I\,(in_1) + \ldots + k_n I\,(in_n) +c$	69.01	4.441e−7
5	$I\,(out) <+\ \texttt{k ddt}(I\,(in_1))$	51.15	4.733e−6
6	$I\,(out) <+\ \texttt{k ddt}(V\,(in_1))$	52.08	4.733e−6
7	$V\,(out) <+\ \texttt{k ddt}(I\,(in_1))$	51.64	4.733e−6
8	$V\,(out) <+\ \texttt{k ddt}(V\,(in_1))$	51.88	4.733e−6
9	$I\,(out) <+\ \texttt{k idt}(I\,(in_1))$	49.01	3.936e−9
10	$I\,(out) <+\ \texttt{k idt}(V\,(in_1))$	48.70	3.936e−9
11	$V\,(out) <+\ \texttt{k idt}(I\,(in_1))$	48.89	3.936e−9
12	$V\,(out) <+\ \texttt{k idt}(V\,(in_1))$	49.16	3.936e−9

Verilog-AMS code matches that needed for the SystemC-AMS simulation. This is due to the fact that both the solvers (i.e., SystemC-AMS and Questa) are solving the same set of equations, as described in Sect. 7.3.

Table 7.2 reports also the level of accuracy *w.r.t.* the expected theoretical results. The error is given in terms of the *Room Mean Square Error*, i.e., by normalizing the error to the mean of the measured values. Thus, it represents the *Coefficient of Variation* between the set of samples gathered during the simulation and the expected theoretical behavior. This proves that the error *w.r.t.* the theoretical results is extremely low, since the Coefficient of Variation between the theoretical reference and the simulation result is always less than $2e - 05$, and in some cases it is as good as $4e - 09$. This error is due to the precision issues of the numerical algorithms used by the simulator to perform continuous time simulation.

The similar simulation times imply that the proposed translation to SystemC-AMS does not provide any simulation speed up, as both the simulators solve the same set of equations with similar strategies. However, Sects. 7.5.4 and 7.5.5 will highlight the effectiveness when handling more complex designs, including mixed analog and discrete descriptions.

7.5.3 Methodology Scalability

In order to show the scalability of the proposed methodology another set of experiments was performed. Two circuits with a single contribution statement were simulated. One circuit has a single non-differential contribution (i.e., $V\,(out) <+ k_1 V\,(in_1) + \ldots + k_n V\,(in_n) +c$), while the second has a differential

Table 7.3 Scalability of the proposed methodology *w.r.t.* the simulation timestep

Input frequency	Adopted timestep	Non-differential		Differential	
		Normalized RMSE	Simulation time (s)	Normalized RMSE	Simulation time (s)
10 Hz	10 ns	4.53e−09	70.35	2.37e−09	51.23
	100 ns	4.53e−08	6.99	5.09e−10	5.09
	1 us	4.53e−07	0.78	1.71e−09	0.52
100 Hz	10 ns	4.44e−08	70.66	2.42e−09	50.98
	100 ns	4.44e−07	7.07	1.24e−09	5.12
	1 us	4.44e−06	0.73	1.66e−07	0.52
1 KHz	10 ns	4.44e−07	70.38	3.94e−09	50.64
	100 ns	4.44e−06	7.07	1.66e−07	5.47
	1 ns	4.44e−05	0.74	1.66e−05	0.52

statement (i.e., $I\,(out) <+ k_1 \text{ddt}\,(V\,(in_1)\,)$). Table 7.3 shows the results of this set of experiments focusing on the two particular kinds of contribution, but similar results apply also to the other case studies. The simulation time refers to the execution of 1 s of simulated time, while the accuracy is given in terms of the Normalized Root Mean Square Error used also for the experiments presented in Table 7.2.

The SystemC-AMS code is stimulated with three different sinusoidal inputs, with increasing maximum input frequencies. For each input, we simulated the code with different time steps, ranging from 10ns up to 1us. The simulation time decreases linearly with the length of the time step, with a speedup of approximately 100× between a timestep of 10ns and a timestep of 1us. At the same time, accuracy is preserved, as the Coefficient of Variation between the theoretical reference and the simulation result is always less than $2e-05$. It is important to note that the error depends both on the adopted timestep, and also on the frequency of the sinusoidal inputs, especially in the non-differential case, where the highest accuracy (i.e., 4.53e-09) is reached with the 10 Hz input and the timestep of 10ns. On the other hand, the combination with the smallest frequency and the largest sample period performs worse than the others, with errors of $4.44e-05$ and $1.66e-05$ for the non-differential and the differential cases, respectively. This is due to the fact that the adopted time step is too coarse for the input frequency. Considering the differential contribution, it worths noticing that once a certain precision is reached, it does not scale linearly as in the non-differential case. However, this happens when an extremely high precision is reached. These considerations highlight the importance of choosing a suitable timestep for the simulation, but also that the generated SystemC-AMS code allows us to determine accuracy/simulation speed trade offs.

Table 7.4 Characteristics of
the original Verilog-AMS
MEMS design

Lines of code		89
Equations	Voltage sources	10
	Current sources	15
Node declarations	Interface	14
	Internal	14
Contributions	Independent	4
	Voltage	59
	Current	0
	Derivative	12
	Integrative	0

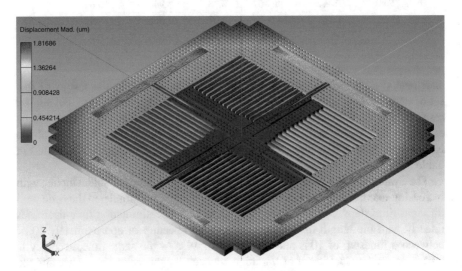

Fig. 7.9 3-dimensional model of the accelerometer in the MEMS+ design simulator

7.5.4 The MEMS Accelerometer

In order to prove the effectiveness of the overall methodology on more complex
designs, we applied the technique to a complex industrial case study, developed
in the context of an industrially-funded project. The case study is a *2-dimensional
MEMS accelerometer* modeled in Verilog-AMS by means of the MEMS design
platform MEMS+. This supports automatic Verilog-AMS code generation [7],
starting from 3-dimensional physical models such as that depicted in Fig. 7.9.
The choice of a MEMS design was guided by the consideration that MEMS
behavioral modeling is based on differential and algebraic equations [8], thus
following the Verilog-AMS structure assumed in this work. Table 7.4 reports the
main characteristics of the MEMS design, both in terms of simultaneous statements
and of types of contributions.

Table 7.5 Characteristics of the generated SystemC-AMS MEMS design

Lines of code		1474
Added node declarations		12
SystemC-AMS primitive instantiations	sca_r	93
	sca_vsource	4
	sca_vcvs	32
	sca_ccvs	0
	sca_csource	0
	sca_vccs	48
	sca_cccs	0
	sca_l	12
	sca_c	0

Table 7.5 shows the results of the application of *ABACuS* to the MEMS design. The table shows the number of lines of code of the resulting SystemC-AMS implementation, the number of added nodes and of instances of SystemC-AMS primitives. The number of lines of codes is increased tenfold (precisely, 11.12×x), as the SystemC-AMS generated by the methodology is more verbose than Verilog-AMS. Each contribution requires the instantiation of the ELN primitive, plus the corresponding explicit port binding. Furthermore, the number of ELN primitives is higher than the number of Verilog-AMS contributions. This is due to the presence of 12 derivative contributions in the original Verilog-AMS code. Each such contribution determines the instantiation of three ELN primitives (as explained in Sect. 7.4.5.3). As a result, of the 188 resulting SystemC-AMS ELN instances:

- 93 correspond to resistors added to connect each SystemC-AMS node to ground;
- 59 correspond to voltage source contributions;
- 36 are generated by the 12 derivative constructs, that also require 12 additional internal nodes.

The numbers highlight that *ABACuS* strictly follows the presented methodology, in particular:

- one resistor is added for each circuit node;
- each non-derivative contribution determines the addition of one ELN primitive instance;
- each derivative contribution generates three ELN primitive instances.

Fast code generation is a major advantage of the proposed approach. Table 7.6 highlights that code generation is almost instantaneous (17.48s overall), and that most of the effort in spent in the HIFSuite conversions (55 %). The most costly step of *ABACuS* execution lies in the mapping from Verilog-AMS contributions to ELN primitives and in their instantiation (37 %). On the other hand, node management and the separation of Verilog-AMS equations into single contributions is almost immediate.

Table 7.6 Characteristics of the execution of *ABACuS* on the MEMS design	Overall		17.48 s
	HIFSuite tools	Conversion to HIF	1.86 s
		Conversion to SystemC-AMS	7.81 s
	ABACuS	Node management	0.94 s
		Division into contributions	0.29 s
		ELN component instantiations	6.58 s

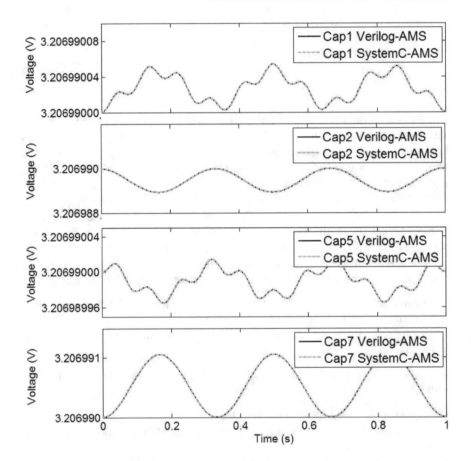

Fig. 7.10 Evolution of the MEMS outputs for Verilog-AMS (*solid*) and SystemC-AMS (*dashed*)

The generated code was validated by comparing its execution *w.r.t.* the original Verilog-AMS code. SystemC-AMS simulation was run by adopting the same input stimula as the Verilog-AMS implementation, and with a 1us timestep. SystemC-AMS simulation proved to be slightly faster than the Verilog-AMS execution (28.02 and 33.72 s, respectively). At the same time, the average error in the computation of the MEMS outputs is 0.02 %. This confirms the visual accuracy evident from Fig. 7.10, where the Verilog-AMS and SystemC-AMS curves are almost totally coincident. The small error is due to the different management of time in the two simulators: SystemC-AMS adopts a fixed timestep, while Verilog-AMS can adapt the length of the timestep over time, thus reaching a higher accuracy. The low error rate highlights the effectiveness of the generated code, both in terms of accuracy and of simulation speed.

7.5.5 Effectiveness of the Proposed Approach

The most important advantage of modeling ABM models in SystemC-AMS lies in the ease of integration in more complex platforms and in the enhanced support for virtual platforms and system-level design, rather than in the pure accuracy or simulation speed.

The SystemC-AMS scheduler is an extension of the discrete-event SystemC scheduler, and it thus allows simultaneous simulation of components belonging to heterogeneous domains. Furthermore, integration with a C++ system level description is eased, thus further removing computationally expensive interfaces and thus speeding up the simulation of mixed-signal systems. For these reasons, SystemC-based languages are a winning solution for the construction and validation of virtual platforms, and they are adopted by most of the currently available virtual platform environments [13–16]. Translating ABM models to SystemC-AMS thus allows their early validation, together with the interaction with other system components.

The MEMS accelerometer has been integrated into a virtual platform for smart systems. The structure of the platform is depicted in Fig. 7.11. It includes (1) a 32-bit RISC processor (the *MIPS CPU*) executing (2) a *Software Application* elaborating data sensed by the accelerometer and stored in (3) a *Memory*. External communication is managed by (4) a Universal Asynchronous Receiver/ Transmitter (*UART*) and by (5) a *Network Interface*, used to send and receive data to and from other smart sensors. All system components, except the MEMS, are modeled in SystemC, and integrated in a virtual platform. Validating the integration of the original MEMS component in the overall system would thus require the construction of a simulation framework.

The first alternative to validate the integration of the MEMS component in the platform is to preserve the language heterogeneity, but within a single simulator. Thus, we adopted Questa [11], which handles both discrete-time and analog descriptions, and that natively provides SPICE-based constructs to connect analog and digital designs.

The second alternative is to adopt the methodology proposed in this work to convert the MEMS design to SystemC-AMS, integrate it in the virtual platform and to run the overall system with the SystemC simulator.

Table 7.7 reports the time needed to simulate 100 ms of real system execution, and it shows how the SystemC based simulation outperforms Questa (by 2.21×). This is mainly due to the heavy communication overhead induced by Questa to allow

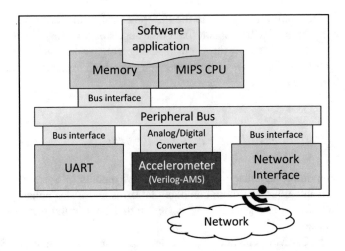

Fig. 7.11 Overview of the virtual platform containing the MEMS (i.e., the Accelerometer) component. Digital components, implemented using SystemC, are colored in *light blue*. The MEMS (*colored in red*) is originally implemented in Verilog-AMS

Table 7.7 Simulation time of the virtual platform by preserving the language heterogeneity and moving to SystemC-AMS

Languages	Simulator	Simulation time (s)
SystemC and Verilog-AMS	Questa	215.47
SystemC and SystemC-AMS	SystemC-AMS kernel	97.59

communication and synchronization between the discrete event and the Spice-base simulators used by Questa respectively for the SystemC and Verilog-AMS parts of the model. At the same time, SystemC-AMS provides a good level of accuracy (0.02 %), thus constituting a valid alternative for early validation of the overall system and of the analog-digital communications.

7.6   Conclusions

The work described here proposes a methodology for representing models that are both conservative and behavioural in SystemC-AMS. We achieve this goal by adopting existing SystemC-AMS ELN primitives in a novel way. As a result, SystemC effectiveness is enhanced in the context of embedded system design, as it can cover a wider range of descriptions and components. Experimental results highlight the correctness of the proposed approach both on synthetic case studies, focusing on the single methodology steps, and on a complex industrial MEMS case study. Future work will focus on the identification of abstraction strategies to target the SystemC-AMS Timed Data Flow (TDF) level for improved simulation performance.

Acknowledgements This work has been partially supported by the European project SMAC FP7-ICT-2011-7-288827.

References

1. IEEE: 1666-2011 - IEEE Standard for Standard SystemC, 2011. standards.ieee.org/findstds/standard/1666-2011.html
2. Zafalon, R.: Smart system design: industrial challenges and perspectives. In: Proceedings of IEEE MDM, p. 3, 2013
3. Accellera Systems Initiative: SystemC-AMS and design of embedded mixed-signal systems, 2013. accellera.org/activities/working-groups/systemc-ams
4. Hartmann, P., Reinkemeier, P., Rettberg, A., Nebel, W.: Modelling control systems in systemC-AMS – benefits and limitations. In: Proceedings of IEEE SOCC, pp. 263–266, 2009
5. Nagel, L.W., Pederson, D.O.: SPICE: Simulation program with integrated circuit emphasis. Electronics Research Laboratory, College of Engineering, University of California (1973)
6. Accellera Systems Initiative: Verilog-AMS, 2014. accellera.org/downloads/standards/v-ams
7. Coventor, Inc: MEMS+: MEMS simulation software, 2013. www.coventor.com/mems-solutions/products/mems
8. Schneider, P., Bayer, C., Einwich, K., Kohler, A.: System level simulation - A core method for efficient design of MEMS and mechatronic systems. In: Proceedings of IEEE SSD, pp. 1–6, 2012
9. Mijalkovic, S.: Advanced circuit and device modeling with Verilog-A. In: Proceedings of IEEE MIEL, pp. 439–442, 2006
10. Narayanan, R., Abbasi, N., Zaki, M., Sammane, G.A., Tahar, S.: On the simulation performance of contemporary AMS hardware description languages. In: Proceedings of IEEE ICM, pp. 361–364, 2008
11. Mentor Graphics: Questa advanced simulator, 2015. www.mentor.com/products/fv/questa
12. Bombieri, N., Di Guglielmo, G., Ferrari, M., Fummi, F., Pravadelli, G., Stefanni, F., Venturelli, A.: Hifsuite: tools for hdl code conversion and manipulation. EURASIP J. Embed. Syst. 2010, 4:1–4:20 (2010)
13. Synopsys: Platform architect, 2016. www.synopsys.com/Prototyping/ArchitectureDesign
14. Cadence: Virtual system platform, 2011. www.cadence.com/products/sd/virtual_system
15. Imperas Software: OVP - Open virtual platforms, 2015. www.ovpworld.org
16. Mentor Graphics: Vista virtual prototyping for SystemC/TLM 2.0 and QEMU, 2015. www.mentor.com/esl/vista/virtual-prototyping

Chapter 8
A System-Level Power Model for AMS-Circuits

Xiao Pan, Javier Moreno Molina, and Christoph Grimm

8.1 Introduction

Renewal of the electronic products, particularly the consumer electronics, is becoming rapider than ever in history. The blowout of the internet-of-things (IoT) and its ecosystems bring this trend more marked. The new product development, in turn, occupies the most important position within the product life cycle. Reduction of time-to-market and development cost are crucial for electronic industrial or companies to hold the trade.

In the design of embedded systems, energy-awareness has set a major trend. Plenty of most recent researches and methods have focused in reducing energy consumption. As a result, ultra-low power architectures with separated power domains have become a common practice. Within these power domains, the supplied voltage can be better optimized, either fixed or using dynamic voltage scaling, or can be even disconnected, when the subsystems that belong to the domain are not in use, using power gating. This segregation into separated power domains has brought about important energy savings, but has also increased the complexity of power distribution.

At the same time, embedded systems have evolved into very heterogeneous systems that integrate digital, analog and RF hardware subsystems, all offering very different sensitivities and vulnerabilities to power integrity problems and contributing to potential cross-talk issues.

X. Pan (✉) • C. Grimm
University of Kaiserslautern, Kaiserslautern, Germany
e-mail: pan@cs.uni-kl.de; grimm@cs.uni-kl.de

J.M. Molina
BBVA, Madrid, Spain
e-mail: javier.moreno.molina@bbva.com

© Springer International Publishing Switzerland 2016
R. Drechsler, R. Wille (eds.), *Languages, Design Methods, and Tools for Electronic System Design*, Lecture Notes in Electrical Engineering 385, DOI 10.1007/978-3-319-31723-6_8

Signal integrity analysis has been traditionally performed very late in the design flow. Indeed, some coupling effects depend on the final circuit layout. However, too late detection makes modifications and fixes very difficult and costly. This chapter aims to provide system-level models that can assist in foreseeing potential power integrity issues that may affect the system at early design stages.

Adding some information about power integrity into the virtual prototyping methodology at an early stage, enables the management of risks and the evaluation of solutions and their effect already during architecture exploration. It can also lead to safer power management strategies and policies that minimize cross-talk.

Furthermore, a system-level model easily permits the evaluation of cross-domain effects. Different subsystems belonging to different domains can be concurrently simulated so that the effects of the aggressor and the verification of the victim can be performed seamlessly.

The next section will discuss the state-of-the-art and previous work. Section 8.3 introduces the concept and the power models. The implementation is described in Sect. 8.4. An use case based on a battery monitoring system is presented in Sect. 8.5, to finally conclude in Sect. 8.6.

8.2 Previous Work

Multi-core architectures and power gating are being increasingly adopted in energy aware designs. However, this introduces additional complexity in the power distribution networks as well as a new source of disturbances in the power supply, as large portions of the system may be switched off or on at unclear moments [13]. In particular the rush current when switching on a subsystem may induce a voltage IR-drop [12]. These power integrity issues can be very critical, as power supply noise may affect different part of the systems. The use of models and simulations can assist in finding this issues at design time.

Modeling coupling effects has been a matter of research for many years. There are many examples of models that aim to detect signal integrity issues. However, most of them address the problem at the circuit level, such as SPICE models and modifications like the one proposed in [11]. There are also some works that study coupling effects on the power supply at the Register-Transfer-Level (RTL) such as those presented in [10] or in [9].

Nevertheless, the increasing complexity of hardware systems demands a higher level of abstraction, not only to simulate those models with the required performance but also to provide some results at early stages in the design, when exploring different solutions is still feasible. A system-level approach to model signal integrity problems in field bus communication using SystemC-AMS has already been proposed in [1, 2]. However, for power integrity such an approach is missing.

This system level approach not only permits an early evaluation of the problem, but will also enable the assessment of power integrity issues in the different subsystems even from different domains. Nowadays, Systems-on-Chip are heterogeneous and include both digital and analog parts. In particular the effect of a noisy power supply is specially critical in some Analog-Mixed Signal components, such as PLLs, where the effect of noise in voltage-controlled oscillators can result into jitter, as shown in [5, 6].

8.3   Power Model: Extended Power State Machine (x-PSM)

Embedded systems consist of a variety of very different components. Besides digital/analog/mixed-signal circuits, such as microprocessors, receivers, transmitters, DAC, ADC; there are many other components consuming power in different ways like electro-magnetic devices, electro-optic devices or electro-mechanical devices that particular widely used in the cyber-physical systems.

A single power model for all such components at a high level would not provide realistic and accurate estimations. For this purpose, we aim at capturing the major components of power consumption, and model them abstractly as follows:

1. Consuming behavior specified by a transfer function. This captures the large signal behavior, e.g. due to charging or discharging capacities etc. when starting a subsystem.
2. A probabilistic model rep. colored noise in the frequency domain, describing the small signal behavior. This captures particular RF behavior of aggressors that introduce spurious tones into sensitive components.
3. A function of input and output signals that describes, e.g. power consumption of class B amplifiers or PWM.

8.3.1   Power State Machine (PSM-) Model

The behavior of components can be modeled at a high level by operation modes that are switched by external inputs. This can be specified by a finite state machine with n states $s_i, i \in \{1, \ldots n\}$. In the PSM-model, the power consumption of each state is described by its average value P_i. Usually, the average power consumption at specific operation states (power states) can be taken from data sheets. Assuming a PSM model, the power consumption at time t is the power at state s_n of PSM would be estimated as constant value

$$P(t) = P_{avg}(s_n) = \Delta E(s_n) / \Delta t(s_n)$$

which is not true, as power consumption is highly variant, and not—as suggested above—piecewise constant with changes only when power states change. However, it is appropriate to estimate energy consumption.

Accounting state transitions is another issue. Depending on the abstraction level, many approaches haven been proposed in the recent years. For peripheral devices modeled with registers (e.g. UART), state transition can be detected by monitoring the contents in the control registers. For the micro-controller modeled at the instruction-level, counting instructions provides enough granularity [8]. State transition for Black-Box IPs can be determined by detecting these activities though the signal that the components exchange [3], or observed by events generated through Protocols (PrPSM) [7].

8.3.2 Extensions to Allow Estimation of Instantaneous Power Consumption

8.3.2.1 Extension 1: Large Signal Model of Power Consumption

Advancements in energy aware and ultra-low power design have led to very aggressive power management techniques, with different power domains that are switched on and off by power gating. Power consumption during state transitions does not change instantaneously. It requires a settling phase to reach a more or less steady state. Often, there are delays, boosting/overshoots, oscillation, glitches or spikes, over the power signals. They are due to the linear behavior of capacities to be loaded and the inductance of wires. In worst cases, there is a risk of suffering a brown-out. However, even very small disturbances can have critical effects.

To capture the large signal behavior of the power consumption at state-changes of the PSM, we assume that delays, overshooting and small oscillations over the power signal can be generally specified by a transfer function $H(s)$. Then the power consumption is (assuming that multiplication of $H(s)$ with a function in time domain is implemented as in most modeling languages):

$$P(t) = H(s) \cdot P_{avg}(s_n) = H(s) \cdot I_{in}(t) \cdot U$$

Note, that $P_{avg}(s_n)$ is a stepwise constant function that changes only when power state change. Measuring the step response of the electric current on the examined component is the simplest way to identify the model, which can be extracted from low level (e.g. RTL) simulations (bottom-up) or measurements on the real hardware (top-down).

8.3.2.2 Extension 2: Small Signal (Noise-) Model of Power Consumption

The power behavior of the AMS circuit is highly dependent on its functionality and internal design. The design of often not known or would introduce a complexity that is inappropriate for system-level modeling.

Therefore, we use a probabilistic approach, assuming that power consumption is a stationary signal when a system is in a power state. We assume that power consumption is colored noise, resp. a random variable (correlated to previous values) in time domain. Furthermore, we assume that the parameters of the noise resp. random process depend on (non-random) parameters, such as operation frequency, supply voltage or input signals that should not be changed for the purpose of characterization of a component.

$$P(s) = H_{noise}(s) \Rightarrow P(t) = H_{noise}(s) \cdot \text{white noise(t)}$$

The noise on the power supply rails $H_{noise}(s)$ of a component can be determined easily via laboratory experiments and measurements, or simulation at circuit-level. Its spectral properties are constant as we assume stationary signals. To determine— e.g. for system simulation—white noise in time domain can be generated and filtered with $H(s)$; again we assume for convenience that the multiplication $\cdot$ of a transfer function with a signal in time domain is implemented as in most simulators.

8.3.2.3   Extension 3: Modulation from an Output or Internal Signal

In most analog circuits power consumption is a function of a signal $s(t)$ that has particular impact, e.g. for amplifiers the power consumption is mostly a function of its output voltage. This can be modeled by a function $f : \{s(t)\} \rightarrow \{P(t)\}$, hence:

$$P(t) = f(s(t))$$

The function $f(t)$ and the specific signal $s(t)$ are often known by a designer based on the known structure of a circuits. For example, the power consumption of an amplified can be modeled as $P(t) = out(t)^2 \cdot const_1 + const_2$, where the constants must be determined by characterization.

The instantaneous power consumption in a power state is then modeled as a linear combination of all three extensions:

$$P(t) = H(s) \cdot P_{avg}(s_n) + H_{noise}(s) \cdot \text{white noise(t)} + f(s(t))$$

8.4   Implementation in SystemC-AMS

The proposed approach has been implemented as part of a simulation framework of networked embedded systems based on SystemC and its extensions for Analog-Mixed Signal (AMS) and communication (TLM). The framework is integrated by a set of libraries with focus in different aspects of embedded systems design. One of these libraries is concerned about energy and power aware design and is named *power-aware framework*.

The power-aware framework was initially motivated to account energy consumption. For that purpose, it contains a power model based on finite state machines where every state is characterized by an average power consumption value.

Although average power is sufficient to obtain energy consumption estimations, it does not allow the required analysis of power integrity problems. Therefore, an extended model capable of reflecting transients is required.

8.4.1 Power Model: x-PSM

The extended power state machine implementation consists of two main elements: states and transitions.

1. **States:** Ordinary states are characterized with a name, an average power consumption value and a vector of possible transitions. The x-PSM maintains the name and the vector of possible transitions but describes the power consumption value as a statistical location-scale distribution. For this purpose three parameters are required: an enum that represents the statistical distribution and two numerical parameters that represent the location and scale parameters.
2. **Transitions:** State transitions are of special interest for modeling power integrity. Section 8.5.1.2 proposed a modeling approach based on Matlab transfer function estimation. For this reason, state transitions are now characterized with previous and next states, the transfer function of the filter that models the transition behavior, and a delay which represents the time specified for transition in the system data-sheet, and that must be consistent with the time required for the signal to stabilize based on the transfer function obtained from measurements.

8.4.2 Power Domains

The stationary effects are provided in the output as statistical parameters and can be computed offline after simulation finishes. The transient effects are simulated using SystemC-AMS timed-data flow (TDF).

All devices connected to a common power supply must be register in a SystemC module called `power_domain`. This module process all the data coming from all the power state machines of the connected devices and outputs the result as a voltage supply TDF signal V_{cc}. This architecture can be seen in Fig. 8.1.

The internal structure of the power domain is shown in Fig. 8.2. Reports are sent automatically when a power state machine changes its state. The `syncer` module is required in order to synchronize the asynchronous state change event with the timed-data flow execution. The `ltf_module` calculates the Laplace-Transfer Function

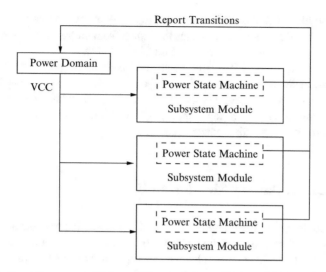

Fig. 8.1 Implementation architecture. State machines report state transitions to the power domain module, which processes them and outputs the effects on an output supply voltage signal

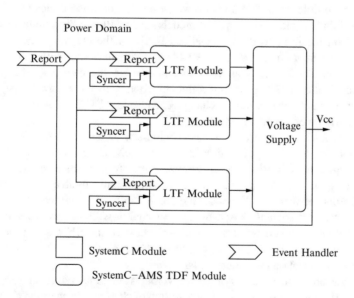

Fig. 8.2 Internal structure of a power domain. Every state machine report is computed by an associated Laplace Transfer Function module. The output of all LTF modules is then calculated in a common voltage supply modules that outputs the V_{CC} signal including the effects of state transitions

(LTF) of the filter with the coefficients used to model the state transition. Finally the `voltage_supply` module combines the inputs from all transitions and generates a voltage supply signal that includes the transient effects.

To generate the model, the user must just instantiate a power domain, which takes as parameter, a vector containing all the modules whose state transitions may affect the power integrity (aggressors). The victims can then take into account those effects by connecting their voltage input port to the output port of the voltage supply module on the corresponding power domain.

8.5 Usecase: Battery Monitoring IC

Our methodology has been applied in a real design of battery management ICs. The project is aimed to build up a framework for design of battery management system (BMS) targeted at system level of abstraction. For this purpose, a building-block library (BBL) that consists basic virtual prototypes, e.g. battery cell, ADC, microprocessor, SPI etc. is created. The designed BMS has to ensure the possibility to handle error and unexpected situations. It is therefore necessary to provoke error situations. In order to enable such investigation, an addition verification library is developed that enables inserting error models such as PI into the simulated system.

Based on the framework, a specialized BMS for the electric vehicle is inves-tigated. The center component of the BMS is the battery monitoring IC (BMIC). As shown in Fig. 8.3, it is an exclusive system on the chip SoC that consists a microcontroller, a SPI slave for communication with the external host, a DDI interface for inner-system data exchanging, and one ADC for monitoring voltage and temperature of the battery cells.

Up to 16 BMICs are wired in a daisy chain network and located under the same power domain, which is powered by the battery cells under monitoring through a voltage regulator. Disturbances on the power supply grid are integrated from all power consumers, as we postulated earlier. Among all possible PIs in the given system, some of them are tolerable and some can be very critical. For instance, the high transient current caused by start of charging/discharging may crash the whole system via a bad isolation (regulator) between the BMICs and the battery cell. Conversely, the noise from the output switching of the ADC through the power grid has less impact on the battery cells.

The measurements accuracy of the cell voltages and temperature are very impor-tant in the design of BMS. It is the key to design of charging/discharging algorithms: mismeasuring may lead to over charging/discharging and in turn reducing the battery lifetime or even cause permanent damaging. The large number of transistors firing at the same time when some components such as the microprocessor are powered on, will have a cumulative effect on the peak current demand. Such current demand by the dynamically switching transistors must be supplied by the

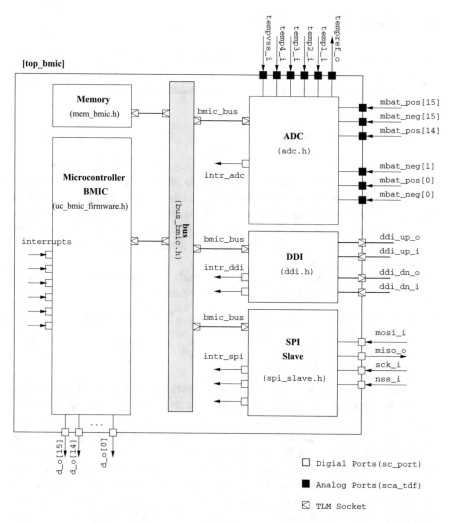

Fig. 8.3 Top level block diagram of the designed battery management IC, which consists a microprocessor, a SPI slave for communication with the external host, a DDI interface for inner-system data exchanging, and one ADC for monitoring voltage and temperature of the battery cells

charge accumulated on these nearby modules or from the regulator, resulting in dynamic voltage drop. In this section we model voltage drop effect on the ADC measurements using the proposed power model: x-PMS. To do so, we first need to develop the power model of the aggressor component, which is the microprocessor in this case.

8.5.1   Power Model Exploration

The power models of the hardware can be characterized from either hardware experiments or circuit level simulation. As such kind of power modeling lacks experiences, focus of this work is on experimental exploration.

8.5.1.1   Experimental Setup

For characterizing the power models, we employ an embedded application for temperature monitoring, in which the components (microprocessor, ADC) are modeled as virtual prototypes in our building block library. The application hardware components are integrated in two separate PCB boards: the MCU board that consists of a microprocessor, and the sensor board that has an ADC with I2C communication interface and a temperature sensor. Figure 8.4 illustrates the architecture of the proposed monitoring system.

The environment monitoring system periodically measures the temperature for a duration of 50 ms. This operation works as described in Fig. 8.6:

1. The microcontroller sends, after powered on, a "start conversion" command to the ADC via I2C.
2. It waits until conversion is done.
3. It reads the converted data from ADC.

Two PCB boards are powered by a high-performance programmable power supply that the output voltage is assumed to be constant according to the specification. The electric current on the power supply line is measured using a hall-effect current probe and recorded by an oscilloscope. In the experiment, we wrap a few wires around the current probe to increase the accuracy and run multiple times to reduce experiment error. Table 8.1 lists the equipment and the ICs used in the experiment.

Power models of the demo system are explored from the measurements of the real hardware (top-down approach). The measurements provide the power consumption behaviors for each hardware component. Note that, the noise coming from the power supply and the oscilloscope is considered to be included in the power models as the extrinsic noise, which is inevitable in the real application.

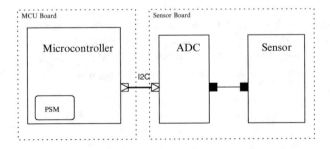

Fig. 8.4 Architecture of the temperature monitoring application for power model exploration

Table 8.1 Hardware and the setups used in the experiment

Equipment/IC	Model	Specifications/settings
Oscilloscope	Tektronic MOS5034	Sampling rate = 500 kS/s
Current probe	Tektronic TCP0030A	Accurately = 1 mA
DC power supply	Hameg HMP2030	Output V = 3.3 V
MCU board	microprocessor	ATSAM3S
Sensor board	ADC	MCP3424
	Temp-sensor	MCP9700A

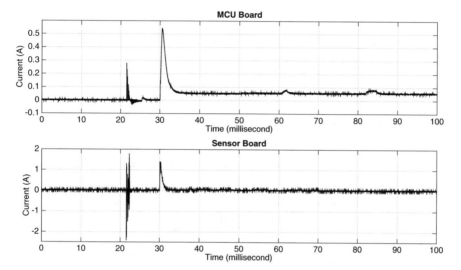

Fig. 8.5 Measured power profiles (rep. in current) of MCU (*top*) and sensor board (*bottom*)

Power consumption behaviors (in measured current, as voltage is constant) from our experiment are shown in Fig. 8.5. The programmable power supply in our experiment has a start-up time to provide stable output voltage (about 30 ms as shown in the figure), in which there can be seen a large oscillations. Since this period is before the experimental system starts up, model of the power behavior is not subject to this work.

8.5.1.2 Power Model Development: X-PMS

During the system activities of TX/RX, the microprocessor and I2C work together, but the power consumption does not significantly vary. The microprocessor in the experiment system has four operation activities and therefore four power states are developed in the PSM model, as shown in Fig. 8.6 (right). Figure 8.7 gives an example of power estimation using PSM model based on the use-case scenario.

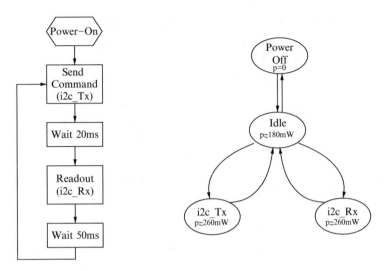

Fig. 8.6 System activities flowchart (*left*) and MCU power state machine (*right*)

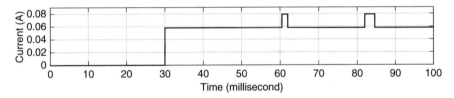

Fig. 8.7 Estimated power consumption (rep. in current) of the microcontroller using power state machine model

Extension 1: Large Signal Model of Power Consumption

We first evaluate the approach for modeling large signal behavior of power consumption caused by state transitions. The inputs are the step signals obtained from the PSM-model of the demo system shown in Fig. 8.7. Therefore, the model parameters are identified from the characteristics of the time domain step response that we logged from the experiment. The power consumption during state transitions is modeled using first-order and higher order transfer function models which are estimated using MATLAB.

The large signal models used to model the power transitions are given in Eq. (8.1) for the state changes between Tx/Rx and idle, and Eq. (8.2) for the state changes between power-on/off and idle, where $I_{(ss)}$ stands for the steady-state current.

$$H_s = I_{(1ss)} \frac{1}{0.00047 \cdot s + 1} \tag{8.1}$$

$$H_s = I_{(2ss)} \frac{24.1 \cdot s^2 + 53340 \cdot s + 4e6}{0.00047 \cdot s^3 + 2.974 \cdot s^2 + 6080 \cdot s + 4e6} \tag{8.2}$$

Extension 2: Signal (Noise) Model of Power Consumption

The noise on the power line is considered a stationary process that produces colored noise. To characterize the model, a spectrum analyzer can be used to capture, in particular, RF behavior; however, for simplicity and proof-of concept we employ the transfer functions from the large signal model for state transitions and $H(s) = 1$ for in power state, multiply by white noise as the small signal model.

8.5.1.3 Comparison of Measurement vs. Large-Signal Model

Figure 8.8 shows the comparison of the estimated power using the large signal model and the measurement. The comparison reveal a very clear output that the large signal modeling approach can simulate the realistic behavior of power consumption at the system level. Note again, the visible deviations before the system start up i.e. the time point at 30 ms in dashed line, come from to the power supply, which are not part of the proposed power model here.

8.5.2 Model Based Simulation

Virtual prototyping has been widely used in the design of embedded systems, which enables faster fully system simulation covering both software and hardware. In our design framework, a building block library with variety virtual prototypes is developed. Considering this work is on the power modeling method and implementation, two associated virtual prototypes in the investigated BMS system—ADC and power supply—will be discussed in a little bit more detail. Power-aware framework which has been presented in Sect. 8.4 is also essentially for the simulation of power models.

8.5.2.1 ADC

The modeling of ADC is based on the existing product which is used for the exploration of power models in Sect. 8.5.1. The ADC module contains the analog and digital parts, implemented in System-AMS and SystemC/TLM respectively, as shown in Fig. 8.9. Input difference between *sca_tdf* ports Vpos and Vneg is read and converted using incremental conversion algorithm into a bit-vector type data. On the other side, a set of mapped registers in the digital part handles the operations of the component.

For the test scenario, the ADC component acts as a victim: fluctuations on the power supply varies the conversion accuracy, i.e. measurements of voltages and temperatures of the battery cells are no longer correct when introducing fluctuations

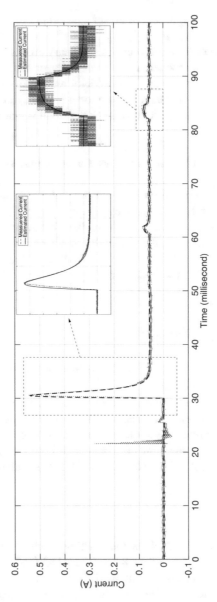

Fig. 8.8 Comparison of measured and estimated power consumption (rep. in current) using the large signal modeling approach

on the power supply. This disturbance to the output is termed as power supply rejection ratio (PSRR) in dB or least significant bits per volt (LSB/V):

$$PSRR = 20log[\frac{\Delta Vdd}{\Delta Vout}]$$
(8.3)

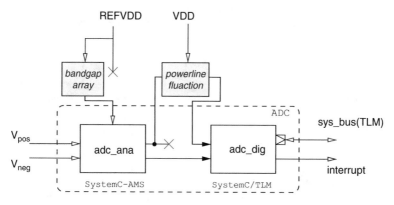

Fig. 8.9 Block diagram of the ADC module with error injection via power supply grid, implemented in SystmC-AMS using TDF MoC and SystemC-TLM. $VDD = 3.3\,V, REFVDD = 1.25\,V, ADCRES = 12\,bits$

Usually, PSRR is given as a stranded parameter in the data sheet. A common value around 40dB is selected to the ADC of the simulation scenario. Note that, the reference voltage error modeling shows in Fig. 8.9 but is not considered here. Because a special error modeling—bandgap error—is used, which is out of the scope of this work.

8.5.2.2   Power Supply

Figure 8.10 shows the schematic of the linear equivalent circuit for the DC power supply, implemented in SystemC-AMS using Electrical Linear Networks (ELN) model of computation (MOC). The supply unit consists of a voltage source Vs (*sca_eln::sca_tdf::sca_vsource*), registers Rdc and Rhf represent impedance at dc and high frequency respectively, an inductance $L0$, and the current source (*sca_eln::sca_tdf::sca_isource*) simulates the current flows through the power supply which is controlled by the power domain unit. The circuit values can be read under the figure. This model suffices for most all regulators (switching or linear) [4].

Usually in the final product, bypassing coupling circuit should be added to reduce voltage fluctuations on the supply line. The proposed power modeling method is also helpful in designing bypassing circuit by verifying through direct simulation.

Power domain unit is the module from power-aware framework that has been introduced in Sect. 8.4. All components containing at least one power consumption model (x-PSM) and powered by the same supply unit are registered to the power domain unit, i.e. microprocessors in this case. Microprocessor module uses the function model that is implemented suing SystemC-TLM. Operation state switching is triggered by the application thread (*sc_thread*) running in the module by calling pre-defined tasks, e.g. SPI-communication, power-off or idle when no tasks running.

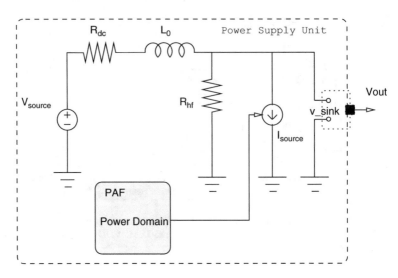

Fig. 8.10 Block diagram of the implementation of the power supply unit (PSU) using SystmC-AMS ELN MoC; xPSM is integrated in power domain (PD) module which controls the current source in the PSU. It is a part of PAF, i.e. power-aware framework, and implemented using in SystmC-AMS TDF MoC; following values are sued in the schematic: $Rdc = 0.02\,Ohms, Rhf = 10\,Ohms, L0 = 90\,uH, Vs = 3.3\,V$

A set of large-signal modes expressed using Laplace transfer function (*sca_tdf* :: *sca_ltf_nd*) are initialized during registration and run in parallel after simulation starts. Power domain unit picks the corresponded LTF module to the output, i.e. value of instantaneous current.

8.5.3 Results and Discussion

Figure 8.11 shows the trace signals of the current and voltage on the power supply grid during state changes of the microcontroller module from $s_{power-off}$ to s_{idle}. As can be read from the figure, maximum voltage drop on the power supply line is -0.2802V versus the supposed supply voltage is 3.3 V. According to the given PSRR definition earlier, we can easily get the measurement error of the ADC: $\Delta V_{Measure} = -0.2802\,V \times 10^{-40/20} = -2.802\,mV$, where the other errors such as bandgap, offset are not count here. Furthermore, this situation can be much worse if multi-microprocessors are powered on/off at the same time. As the minimum requirement of the measurement error of the battery voltage in this project is 3 mV, the proposed design of the power supply unit is not good enough for our battery monitoring IC. Improvement such as bypassing circuit has to be added.

Some strategies can be used to reduce the disturbances depend on the field of the application. For example the hardware engineer can make better power supply unit

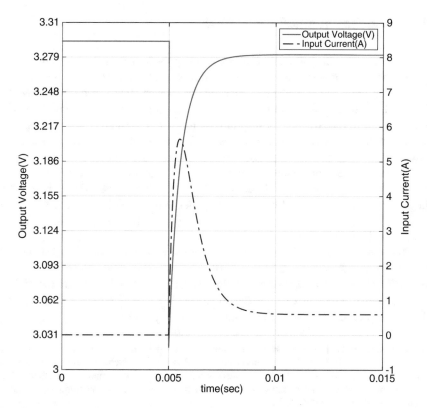

Fig. 8.11 Simulation output of power supply voltage and current in state transitions of the microcontroller from $s_{power-off}$ to s_{idle}. (without bypassing circuit)

by adding proper bypassing circuit. However on the other side, adequate bypassing requires big capacity, which is usually large to the integrated circuit and causes circuit backflow and then more ripples on the power supply line. Alternatively software engineers must develop error-aware charging/discharging algorithm.

Although the simulation performance is not the major concern of this work, it is still meaningful to have a discussion about it. As can be seen from the description of the implementation of the power model, one Laplace transfer function (*sca_tdf::sca_ltf_nd*) is instantiated for each state transition existed in the system, and runs continuously till the end of the simulation. That is because the solver of the Systemc-AMS for *sca_tdf::sca_ltf_nd* does not allow changing the order after simulation starts and unique order transfer function could not represent varied behaviors of the large signal models. Furthermore smaller time-step is necessary to achieve higher precise or correct computations. These are big challenges to outcome high speed simulation till the present work progress.

8.6 Conclusion and Future Work

In this chapter we proposed and evaluated an approach that for the first time allows to estimate instantaneous power consumption of AMS circuits at a high level of abstraction. The modeling approach extended power state machine (x-PSM) is intended to permit analysis of the power distribution network and its infrastructure (e.g. DC-DC converters) and its dimensioning. Furthermore, in particular the extension of small signal power behavior in frequency domain is useful to analyze cross-coupling via power supply rails in a very early development stage.

The parameters for modeling small and large signal power behavior are easy to determine by experimental measurements or circuit-level simulations. The large signal behavior model has been integrated into a simulation framework and evaluated on a real design. Implementation of the small signal behavior model is subject of the future work.

The experimental results in the use case of the battery monitoring IC show that the modeling method is accurate and allows functional validation of the power-integrity at the system level of abstraction. Where it is, as we postulated, useful to show potential issues in a design has to be shown by more complex case studies. This is subject of future work, based on the concept presented here.

Acknowledgements The work presented in this chapter has been carried out in the ANCONA project, funded by BMBF (Bundesministerium für Bildung und Forschung) program IKT 2020 under contract no. 16ES021. The case study system is conducted as part of the "Integrierte Komponenten und integrierter Entwurf energie- effizienter Batteriesysteme (IKEBA)" project, supported by BMBF program IKT 2020.

References

1. Alassir, M., Denoulet, J., Romain, O., Suissa, A., Garda, P.: Modelling field bus communications in mixed-signal embedded systems. EURASIP J. Embed. Syst. **2008**, 1 (2008)
2. Alassir, M.D., Denoulet, J., Romain, O., Garda, P.: Signal integrity-aware virtual prototyping of field bus-based embedded systems. IEEE Trans. Compon. Packag. Manuf. Technol. **3**(12), 2081–2091 (2013)
3. Atitallah, R.B., Meftali, S., Dekeyser, J.-L., Trabelsi, C., Jemai, Abderrazek.: A model-driven approach for hybrid power estimation in embedded systems design. EURASIP J. Embed. Syst. **2011**(1), 15 (2011). Article ID 569031
4. Calex.: Understanding power impedance supply for optimum decoupling. Feb 2008
5. Heydari, P.: Analysis of the PLL jitter due to power/ground and substrate noise. IEEE Trans. Circuits Syst. Regul. Pap. **51**(12), 2404–2416 (2004)
6. Heydari, P., Pedram, M.: Analysis of jitter due to power-supply noise in phase-locked loops. In: Proceedings of the IEEE 2000 Custom Integrated Circuits Conference, 2000. CICC, pp. 443–446. IEEE, New York (2000)
7. Lorenz, D., Hartmann, P., Gruettner, K., Nebel, W.: Non-invasive power simulation at system-level with systemC. In: Ayala, J., Shang, D., Yakovlev, A. (eds.) Integrated Circuit and System Design. Power and Timing Modeling, Optimization and Simulation. Lecture Notes in Computer Science, vol. 7606, pp. 21–31. Springer, Berlin (2013)

8. Molina, J.M., Haase, J., Grimm, C.: Energy consumption estimation and profiling in wireless sensor networks. In: 2010 23rd International Conference on Architecture of Computing Systems (ARCS), pp. 1–6. VDE (2010)
9. Schumacher, P., Jha, P., Kuntur, S., Burke, T., Frost, A.: Fast RTL power estimation for FPGA designs. In: 2011 International Conference on Field Programmable Logic and Applications (FPL), pp. 343–348, Sept 2011
10. van Heijningen, M., Badaroglu, Mustafa., Donnay, S., Engels, M., Bolsens, I.: High-level simulation of substrate noise generation including power supply noise coupling. In: Proceedings of the 37th Annual Design Automation Conference, DAC '00, pp. 446–451. ACM, New York, NY (2000)
11. Verghese, N.K., Allstot, D.J., Masui, S.: Rapid simulation of substrate coupling effects in mixed-mode ICs. In: Proceedings of the IEEE 1993 Custom Integrated Circuits Conference, 1993, pp. 18.3.1–18.3.4, May 1993
12. Xu, T., Li, P., Yan, B.: Decoupling for power gating: Sources of power noise and design strategies. In: 2011 48th ACM/EDAC/IEEE Design Automation Conference (DAC), pp. 1002–1007, June 2011
13. Zeng, Z., Feng, Z., Li, P.: Efficient checking of power delivery integrity for power gating. In: 2011, 12th International Symposium on Quality Electronic Design (ISQED), pp. 1–8, Mar 2011

Printed in the United States
By Bookmasters